Image Control

Image Control

Motion Picture and Video Camera Filters and Lab Techniques

Gerald Hirschfeld, A.S.C.

Edited by Julia Tucker

Focal Press

Boston London

Cover photograph: Grainger House, Ashland, Oregon.

Cover photo filter information left to right:
1. No filter
2. Sky control (Chapter 4, Graduated Filters)
3. Sepia effect (Chapter 4, Coral Filters)
4. Day-for-night (Chapter 5, Creating Day-for-Night Effect)
5. Fantasy effect (Chapter 4, Fog Filters)

Focal Press is an imprint of Butterworth–Heinemann.

Library of Congress Cataloging-in-Publication Data
Hirschfeld, Gerald.
Image control: motion picture and video camera filters and lab techniques/ Gerald Hirschfeld: edited by Julia Tucker.
p. cm.
Includes index.
ISBN 0-240-80167-9 (alk. paper): $49.95
1. Cinematography. 2. Motion picture cameras. 3. Photography—Camera filters. 4. Cinematography—Processing.
I. Tucker, Julia. II. Title.
TR850.H549 1993
778.5'3—dc20 92-24665
CIP

British Library Cataloguing-in-Publication Data
A catalogue record for this book is available from the British Library.

Butterworth–Heinemann
80 Montvale Avenue
Stoneham, MA 02180

10 9 8 7 6 5 4 3 2 1

Printed in the United States of America

This book is dedicated to my sister, Herminie,
who gave me a home developing kit for my fourteenth birthday
and helped me find my first professional job as a darkroom assistant,
and,
To my mother, Katherine,
who gave me continuous encouragement
in all my photographic endeavors
and saw only great pictures emerge from my camera.

Contents

Foreword

My first film as an actor was *Goodbye, Columbus*. It was a thoroughly enjoyable experience— cast and crew became one during the shooting. Some of us were in front of the camera, some behind, all dedicated to making a good film, all supporting each other. You go into a film strangers and emerge at the end a tight combat unit complete with war stories to tell your grandchildren. The only difference in making a movie and combat is that no one is shooting at you—come to think of it, sometimes they are.

Out of that film came a lasting friendship with the picture's director of photography, Gerald Hirschfeld. I found in Jerry the special qualities of the best cinematographers: technical expertise coupled with emotion and passion and, in Jerry's case, a kindness and a caring.

When it was time to direct my first film, *My Favorite Year*, the first thought I had was who will be the cameraman. For me, Jerry, of course, was the obvious choice. Once again, the experience was a total pleasure, but on another level. Jerry helped me when I needed it, allowed me to find my way without embarrassment, and finally, did the impossible when someone else might have said it couldn't be done.

One night, I was shooting a scene in front of NBC at Rockefeller Center with Peter O'Toole and Mark Linn-Baker. When we discussed the shooting of the scene days before, Jerry asked me if we would be seeing three blocks south during one of the shots. I said no. Naturally, on the night of shooting, I changed my mind. Jerry said, "We don't have the equipment to light three blocks of New York real estate." He paused, "But give me a few minutes." Somehow, by his talent, he made it work with what he had, and it was beautiful.

I am honored to be asked to do the foreword to this book. Keep in mind as you explore the tools of technique that everything is really emotional— that these tools, in the right hands, illuminate the human condition. No one does it better than Gerald Hirschfeld.

Richard Benjamin

Preface

This is a nontechnical technical book. It's nontechnical in the sense that it doesn't deal with theories or formulas; it's technical in that it gives the practical, visual results of what happens to a color negative when certain techniques are applied. You might be interested to know how this book came into being. For the 40-odd years that I've been making motion pictures, I've made film tests that are quite similar to those described in this book. It was the only way to discover exactly what laboratory processes did to a color negative and to learn exactly what the results of using a particular filter would be. Tests of this nature are not unique by any means; most professional cinematographers shoot tests very frequently, especially when they are working with a new color emulsion or planning a special look for a new film.

Why hasn't a book like this appeared before now? I think I can safely say that many cinematographers have never taken the time to write down the information they have learned from their years of experience. I know that for a long time I never did. Then something happened in the summer of 1984 that gave me a fresh perspective.

An unexpected telephone call from the Tahoe Film and Video Workshop brought an invitation to conduct a two-week seminar on cinematography for a select group of aspiring filmmakers. With such a short time for the seminar, I asked the participants which phase of cinematography interested them the most. The answer was the various aspects of controlling the look of a film. This covered a wide range of topics, including lighting, lab techniques, color negative emulsions, and of course camera filters, which was the subject we discussed primarily. At the end of the seminar, I realized that I had been able to communicate to these young people some of the knowledge from my experience as a director of photography on more than 40 feature films, countless television commercials, and hundreds of film tests. My seminar offered them insights into cinematography that are not often available from other film classes and gave them the point of view of a working professional. Their enthusiasm and acceptance, their eagerness to learn, and their appreciation of my accomplishments gave me an inner satisfaction that was very meaningful. The response to sharing my knowledge coupled with my desire to prevent valuable information from fading away gave rise to the thoughts that have culminated in this book. I realized that it was time to pass along the essence of my experience.

This book is directed toward all those who have the ambition to become the very best cinematographers and videographers they can be. It contains information for all camera and video persons, be they documentary, travel, or nature filmmakers; television commercial or feature film directors of photography; serious amateurs; beginning professionals; or still photographers.

Enjoy this book. Use it to enhance your own experience in this most imaginative medium of communication: the visual image. Let it serve as a constant stimulus to your own thoughts and visions.

NOTES ON THE ILLUSTRATIONS

Every effort has been made to match the color reproductions in this book as closely as possible to the original print from the color negative. However, even though these reproductions are made on high-quality printing presses, they are pigment images impressed on paper and, as such, can never exactly match the quality of a projected print.

To present a format larger than the single frame of a motion picture negative (so that the details discussed would be more visible), most of the examples were filmed as double-frame 35mm stills. Nearly all the examples of filtration and processing were filmed on either Eastman Kodak #5247 or #5296 and processed at RGB Color Lab in Hollywood or at Evergreen Film Service in Eugene, Oregon.

When a series of filters is discussed, it is impossible to demonstrate through a paper reproduction the subtleties of change with each degree of filtration. Therefore, only those examples that show a distinctive change are used, and the reader must interpolate the missing images.

Many of the illustrations are scenes without people. People can usually be lighted easily, whereas the background area or landscape can often be controlled only through the techniques described. When a person is in the scene and properly lit, the effect of adding a filter can be seen on the ground glass of the camera. It's the director of photography's job to make sure that the scenes that include people match the longer shots where people are either not present or so small that their presence does not affect the filter selection.

NOTES ON COMPANIES AND PRODUCTS

Throughout this book, various manufacturers of filters and color negatives and various film processing labs are discussed. Specific manufacturers are mentioned because they are the ones that I am familiar with. For example, in 1941 the only filter manufacturers for the professional film business that I was aware of were Eastman Kodak, Harrison & Harrison, and a specialty filter maker, Scheibe. The variety of available filters was limited. As the demand for filters increased, Tiffen and Harrison & Harrison became the main producers of filters for professional filming. These are the manufacturers that I "grew up" with. However, other manufacturers also make excellent filters, and different cinematographers have personal preferences for the filters they use, just as they prefer different film stocks and different film processing labs. I have only worked with Eastman and Fuji film stocks, but these films are certainly not superior to Agfa. Films that have won Academy Awards for cinematography have been shot on all three negative stocks. The reader should not misconstrue the mention of a particular manufacturer or product as an endorsement of only that company or that product.

Acknowledgments

Writing a book is much like making a theatrical film: it's never a one-person creation. Much of the prompting to set down this material came from camera operators and assistants who worked with me on various feature films and wanted to know the reasons for my using particular filters or lab techniques.

To the following companies and directors of photography, I give my heartfelt thanks for their help and contributions to this project.

American Society of Cinematographers
Arriflex Corp.
Cokin Filter System
Consolidated Film Industries
Color Magic Filters, Inc.
Eastman Kodak Company
Evergreen Film Service
Fuji Photo Film Corp.
Panavision Corp.
Pat Perlson and Fred Gledhill
Harrison & Harrison Filters
Pancro-Mirror, Inc.
Sandler Films
Spectratek Professional Camera Filters
Sylvia Beach Hotel, Newport, Oregon
Technicolor Laboratory
Tiffen Filter Company

Nestor Almendros, A.S.C.
John A. Alonzo, A.S.C.
Owen Roizman, A.S.C.
Giuseppe Rotunno, A.I.C., A.S.C.
Vittorio Storaro, A.I.C., A.S.C.
Alex Thomson, B.S.C.
Zoli Vidor, A.S.C.
Vilmos Zsigmond, A.S.C.

Special appreciation goes to Julia Tucker, who helped with everything. She understood what I wanted to say and therefore was able to revise my prose as well as edit the manuscript in the usual sense. Thank you, Julia—my partner, my best friend, my wife.

Image Control

Chapter 1
Reasons for Altering the Original Color Negative

Why do anything to alter the color negative when film manufacturers have succeeded in creating emulsions that accurately reproduce all the colors of nature? You might also ask why film laboratories have a variety of processing techniques or why filter manufacturers have created hundreds of filters. The answers are complex, but this is a beginning thought: The realistic reproduction of color is objective, but creating a motion picture is subjective; the latter often requires that reality be altered to make a complete dramatic statement.

VISUALLY ENRICHING THE STORY

A motion picture's prime purpose is to tell a story. It follows that different types of stories require different "looks." The look of a film can greatly enrich the audience's reaction.

If the screenplay communicates impending danger through dialogue and stage direction, the photography can support that mood through low-key, shadowy lighting. Even in day exterior scenes, the style of photography can create a mysterious mood. On the other hand, if the screenplay is comedic or light, the photography can be bright, pastel, and happy in feeling. Between these two extremes, there's a whole world of possibilities, and the cinematographer has a wide range of styles from which to choose.

Photographic adages used to dictate that the sun come from over your shoulder and that the ratio of key light to fill light be about three to one. Today, however, hard-and-fast rules have given way to individual interpretation, and the original color negative can be changed to reflect various moods to better express the intention of the screenplay.

SUBLIMINALLY ENHANCING THE SCENE

The style of lighting, which helps create the mood of a film, is the strongest visual statement that can affect the viewer and give the film a specific look. Lighting is a complex subject in itself. What I'm touching on here is the enhancement of the lighting through a subtle change in the image, something that's incorporated in the screen image but isn't at all obvious because it operates below the threshold of consciousness. It's a subliminal effect. Various filters and laboratory techniques can do wonders to create many subliminal effects. They can intensify the mood of a scene, lend a diversity of looks, and most importantly evoke a particular response from the viewer.

You may recall some isolated film sequences that had a look different from that of the rest of the film. Dream sequences are often vignetted and diffused. Color films often have flashback sequences in sepia tone or even in black and white. Sequences involving a particular character may be distinguished by using a special filter combination created specifically for that character, and these filters may change in the course of the story as the character's circumstances change from, say, a life of struggle to a life of ease. In an otherwise normal-looking film, a location such as a bar, night-club, or discotheque can have an altogether different look. A lyrical look is often created by using diffusion, muted colors, and perhaps a bit of fog to produce a feeling of fantasy.

These kinds of scene enhancements are used to evoke a particular subliminal reaction from the audience. To contribute graphically to the interpretation of the screenplay is one of the most sensitive and satisfying responsibilities of the director of photography.

ACCOMMODATING POST-PRODUCTION WORK

Often, on completion of the principal photography of a theatrical film, only part of the film's final look has been achieved. With the sophistication of today's filmmaking, post-production work on the negative plays a very important role, and the cinematographer can help greatly with the final results.

In relation to the original color negative, *post production* refers to the work to be done with the negative after the principal, or main, photography has been completed. At this time, optical effects such as fades, dissolves, and double exposures are added. *Post production* can also refer to matte work in which part of the scene is removed and replaced by artwork or a miniature or, in the case of a blue screen matte shot, the actors are placed against an entirely new background. Post production may simply mean adding titles or subtitles for dialogue translation. There are other post-production additions, but they need not be considered in the context of this book.

The medium for which a film is primarily made is important. Is it being produced for 35mm or 70mm release in theaters, or, like a commercial, is it being made specifically for television? Will it be released in 16mm or as a videocassette?

The theatrical film release represents the purest form of viewing the original film. Even in this medium, you must be aware of any special work to be done in post production that could adversely affect the look of the film. As an example, if you are aware that a specific scene is to be used in a composite matte, it is best to shoot the original scene without special filters. You don't want to combine a filtered shot with an unfiltered shot in a composite scene. Additional opticals or lab techniques can match the newly created composite scene to neighboring scenes in the same sequence that may have been filtered.

A film or commercial made specifically for television faces certain restrictions imposed by the television transmission system. The tonal range, or contrast, of these films must be kept within the limits that can be reproduced most effectively. Eastman has helped tremendously in reducing normal print contrast by introducing low-contrast print stock (#5380 for 35mm and #7380 for 16mm) specifically designed for making prints for film-to-tape transfers.

Additional information on post production is included in Chapter 6.

DECEIVING THE EYE

Film manufacturers achieve their goal when their product is able to reproduce exactly the colors of nature. As mentioned earlier, they have done just that. If you were to shoot a landscape on a sunny day using only normal procedures to make the color negative and its print, and then you looked at that print at exactly the same location under the same conditions, you would probably agree that the blue sky, the green trees, and the golden field of wheat were accurately reproduced. The film exactly duplicated the colors of nature. However, take that same print, project it in a darkened theater, and surround the screen with a large black border, and you will probably feel that the sky is too blue, the trees a bit too green, and the wheat too golden. Why?

There are several situations in which the eye can be deceived. First, by surrounding the image with black, we immediately heighten the apparent color saturation of the scene because we have removed the portion of the light that normally enters through our peripheral vision. It is this light that tends to slightly mute the colors. To test for this effect, take a 5" × 7" normal color picture, surround it with a very wide black border, and on top of that place a white border of the same size. With plenty of light, keep your eyes focused on the colors of the picture with the white border in place, then quickly whisk it away and look at the picture surrounded by the black border. The colors will probably appear to be more intense with the black border than with the white one. The white border reflects some white light into your eyes, which slightly mutes the color. When replaced by the black border, the slight flare of the white border is eliminated, and the colors appear more saturated.

Another deception occurs outdoors where the array of colors we see is relatively boundless in scope and our eyes look only at the red barn or selectively concentrate on the green trees or the blue sky. While looking at the barn, we are not simultaneously as aware of the trees or the sky; that is, objects outside the area of our concentration. However, when this expansive scene is reduced to the size of a motion picture screen, the eye is able to see adjacent colors almost simultaneously. When the eye responds to the red, green, and blue virtually at the same time, the interaction of color makes them seem more saturated, particularly when surrounded by the blackness of the theatre.

Strange as this may seem, another result of the interaction of color is the apparent change of one color when adjacent to another color. The most thorough and explicit book on this subject is *The Interaction of Color* by Professor Josef Albers, (New Haven: Yale University Press, 1963). The book has a large format text of only 80 pages, but it is accompanied by a three-inch-thick folder containing color plates and other samples to illustrate the way colors interact with one another and how our eyes are deceived by that interaction.

Albers writes: "In visual perception a color is almost never seen as it physically is. This fact makes color the most relative medium in art. In order to use color effectively it is necessary to recognize that color deceives continually."

Albers explains that the differences between color deception and color perception may alter the actual visual effect of a color. In simplest terms, not everyone sees colors in the same way. In an extreme case, I once worked with a director who was color blind; although he liked what he saw on the screen, I had no way of knowing how my color photography looked to him.

Albers writes: "If one says 'red' (the name of a color) and there are 50 people listening, it can be expected that there will be 50 reds in their minds. And one can be sure that all these reds will be different. Even when a certain color is specified, which all the listeners have seen innumerable times, such as the red of Coca Cola signs, which is the same red all over the country, they will still think of many different reds. Even if all the listeners have hundreds of reds in front of them from which to choose the Coca Cola red, they will again select quite different colors. And, no one can be sure that he has found the precise red shade....Each will receive the same projection on their retina, but no one can be sure whether each has the same perception."

Awareness of the phenomenon of color interaction and color deception is vital to the cinematographer. Reproduced in Color Plate 1 are examples from *The Interaction of Color* that graphically demonstrate how the eye is fooled or visually misdirected. Even these few examples show how the interaction of color can alter the appearance of a color or shading of a subject. You never know when a situation will arise in which one color, because of where it is in relation to another color, will appear different. Not only is there an interaction between colors, there can also be an interaction between a color and gray. It's the adjacent area to gray that may affect the apparent shade of gray. You must scrutinize the colors of a scene to prevent something other than what was intended from being created.

If you were to film people with light flesh tones against a black backdrop, they would seem much paler than normal; conversely, those same people in front of a white backdrop would seem a bit tanned. Precise laboratory printing of these two situations can correct the deception by printing figures darker when against a black background and lighter when against a white background. Think of a snow or beach scene you have shot. Don't the people you photographed seem to have ruddier complexions than normal? This is more an interaction of the background brightness than of color, but the principles are similar. Albers confirms that the eye of the beholder is subject to a variety of false impressions. (See Figure 1.)

A quite different effect, psychological in nature, is created when the brain is programmed to help supply a color in a situation where the actual color is not perceived by the eye. When you are confronted with colors and fabric textures that are markedly different from normal in color saturation and reflective values, it can be quite a challenge to reproduce them faithfully. Many are highly reflective; for example, lighter colors and smooth, shiny cloth. Other colors such as dark burgundy, forest green, or navy blue and fabrics like deep velour, velvet, or thickly textured wool have little reflective ability and must be specially lighted if they are to be reproduced accurately on film. Obviously, you cannot continually throw extra light on the actors' wardrobe as they move around the set! In these instances, you receive tremendous help from the retentive power of the brain. If you can establish the problem color early in the sequence, the brain will continue to remember that color almost exactly. For example, if an actor is wearing a dark burgundy velvet jacket, it behooves you to get plenty of light on it early on to establish the color and texture, assuming that this wardrobe is important to the definition of the character. If the actor is seated near a light source such as a lamp or a sunlit window, there is an opportunity to throw a lot of light on the fabric (without spilling any extra light on the actor's face). In the rest of that sequence, when the jacket is barely lighted, the brain will supply to the image, the color and texture information that the eye does not actually see.

Very light fabrics create a different problem, but the brain can again supply information that is perceived differently through the eye. As an example, imagine an actor wearing a white jacket; the color is conveyed accurately through normal lighting. Then in the following sequence, the actor appears in a night exterior scene where you are using a bluish moonlight and underexposing to enhance the effect. The white jacket now appears to be blue; however, the brain has already acknowledged it as white, and the original information supersedes the new information. The audience knows that the actor is still wearing the white jacket and hasn't changed into a blue one.

In summary, the eye and the brain work together to determine how colors are perceived. It is the director of photography's job to make that perception accurate for the audience and to reproduce authentic colors on the screen. If a color looks abnormal due to its interaction with an adjacent color, believe what your eyes tell you. Even though the color film records all colors accurately, the eye reacts similarly to the color reproduction as it would to the original

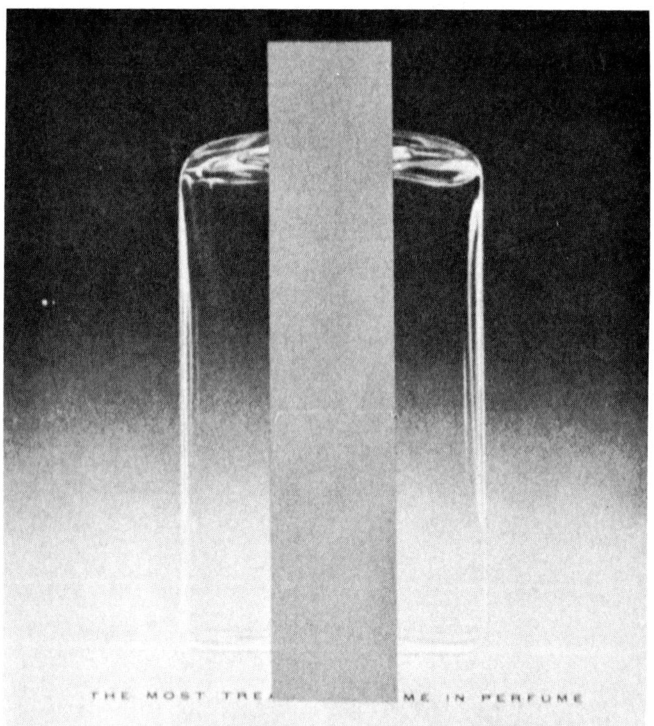

Figure 1 The gray vertical bar appears to be darker at the bottom and lighter at the top. Place white index cards on both sides of the gray bar and you'll see that it's one continuous gray strip. The illusion is caused by interaction with the background shading. Study the subject in Color Plates **1E** and **1F**; her hair, face, and sweater appear lighter against the black background than against the white one. By masking everything but her face with a neutral covering, thus eliminating the background, you'll see that the brightness of the subject is exactly the same in both situations.

color if there is in fact an interaction with an adjacent color. In other words, acknowledge the deception and compensate for it so that the final result is what you want the viewer to see. Also, remember that the viewer will supply remembered color and texture in certain circumstances. Although the colors of nature are accurately reproduced, the visual impression in the dark theater is one of increased color saturation. These are some of the reasons to alter the color response of the color negative. To achieve exactly the result that you want it is often necessary to manipulate the image. The techniques for doing that are described in this book.

COMPENSATING FOR FILM LABORATORY VARIABLES

The film laboratory plays an important role not only in post production but also in developing and printing the negative for dailies. An original color negative is almost never used to make release prints. It must be protected, so a duplicate negative is made, which may affect contrast and color saturation. In addition, certain scenes may include optical effects, which may also affect contrast and color saturation. Later chapters explore the options that the director of photography has to control these situations.

Film manufacturers and film-lab personnel might object to this statement, but it's a fact that the duplicate negative, or dupe, does not always match the original. Inconsistencies in developing the dupe negative can make it extremely difficult to produce a perfect match each time.

If you have had film processed in different labs you may have seen variations in the areas of contrast and grain. Such dissimilarities should not occur because all labs have the same processing information with respect to formulas, time, and temperature, but even the water used in the chemical baths can make a difference, water filters not withstanding! I had a color negative developed and printed in England, and an American lab could not match the print from the very same negative. It was ultimately discovered that the difference was the water used in the processing baths!

The film lab is one of the elements that may affect the final image on the screen. Knowing in advance the degree of variance from the norm that a particular lab provides, enables you to take certain steps to help compensate for this variance when you start filming.

As you work with film you'll probably find many instances in which you'll want to alter the normal color response of the color negative.

Chapter 2
Controlling the Color Response of the Color Negative

For this discussion, let's assume that the term *color response* means not only the color that the color negative produces but also anything that can affect it, for example, negative density, color saturation, color contrast, graininess, and of course color correction and color shifting. If any of these terms are unfamiliar, they can be found in the Glossary.

PRODUCTION DESIGNER, COSTUME DESIGNER, AND SET DECORATOR

Besides camera and laboratory techniques, cinematographers can affect the color response of the color negative through the cooperation of three important production people: the production designer, costume designer, and set decorator. The color of a wall greatly affects not only the scene itself, but also any subject in front of the wall. An actress wearing a red dress in front of a blue wall creates a color interaction as well as great color contrast. If the actress wears a shade of blue that complements the blue wall, a different and very becoming effect would be created. If the wall is a warm gray, the scene may be much more pleasing in terms of color contrast. I don't think I'm alone when I say that my favorite wall colors are shades of gray or off-white; they're most flattering to flesh tones and eliminate any interaction of color. Similarly, furniture chosen by the set decorator can either create or avoid problems for wardrobe colors and other colors. Set colors, costumes, and patterned materials all have great importance in the overall look of a film.

Many films that are "down" in subject matter rather than "up" feature earth tones, for example, brown, gray, beige, umber, wheat, and green. These colors greatly affect the overall impression created for the film. Against this drab palette, brightly colored costumes or objects such as cars can become strong accents. Again, the interaction of color. Designers and set decorators make significant contributions to a film, so diplomacy is key in your communication with them. Discussions with the director concerning the look do much to clarify the situation.

Don't be shy about sharing your ideas regarding the look you want to create; the elements that other departments contribute make your work look that much better. Actors have enormous influence over what they wear, so it might be wise to let the director guide their choices so that the wardrobe will support not only the character, but also the look of the film.

If the areas of set design, wardrobe, and set decorating interest you, it would be of great help to look at design books on these subjects. Discussions with professionals in these fields are helpful too. Another rich source of information resides in films themselves—films that benefited from major contributions from the professionals in these crafts. Videotape rental stores offer a full library of films for study. Explore the films that won Academy Awards for production design, wardrobe, and set decoration.

It's the cinematographer's expertise that unifies all of these elements: enhancing design with lighting and other visual effects, establishing a special look that helps tell the story, and creating a visual ambiance that allows the audience to enter the experience of the film. Without a close working relationship with others, however, much of what the cinematographer envisions can remain unrealized.

THE EFFECTIVE CONTROL OF LIGHT

Lighting is the predominant ingredient in the control of the film image. Many books have been published exclusively on the subject of light and its application. The following discussion by no means covers the whole subject, but certain points are included here because light so strongly affects the color response of the color negative.

I'm sure you know that some of the most beautiful exterior scenes are those filmed in bright overcast or hazy sunlight. This condition is created by a high, thin cloud layer or heavy atmospheric haze, and it is equivalent to setting up a gigantic silk diffuser to soften the direct sunlight. Overcast light brightens the dark, usually objectionable shadows caused by harsh clear sunlight. With bright overcast, even if the sun is directly overhead (photographically, the least desirable place for it to be), the shadows are sufficiently softened; with a minimum of fill light, a pleasing picture results. If a hazy sun is low in the sky, a beautiful, softened light wraps around the subject. You would be extremely fortunate to have this light condition repeat itself for the time required to complete a sequence, which can be several days! However, with the cooperation of the director, all the long shots can be made under the bright overcast condition. It's then possible, if the sun comes out, to control the light for the closer shots by using overhead silks and nets. This optimum soft light creates a relatively low ratio of highlight-to-shadow brightness, which is very becoming for many films.

If you have direct sunlight on your subject and the size of the scene makes it impossible to rig an overhead silk, one way to reduce the contrast ratio is to use fill light to brighten the shadow areas and thereby reduce the extreme highlight-to-shadow ratio. Any of the following, in any combination, can be used for fill: an arc light, HMI light, reflectors, a griffolyn, and bounce boards. Indoors, where the cinematographer has complete control over the lighting, any kind of light—from a direct spot to a soft bounce light—can be used with as much fill light as necessary.

When working outdoors, the director of photography controls the lighting by choosing the time of day a particular scene will be filmed. Additional control comes from placing the camera relative to the position of the sun. Even with absolutely no lighting equipment, the cinematographer can get a pleasing lighting effect outdoors in bright sunlight simply by placing the camera so that the sunlight becomes a back light and setting the exposure for shadow detail. This creates a much softer image than when the sun strikes the actor's face directly, causing harsh shadows and uncontrollable squinting. For those scenes that would benefit from a more dramatic feeling, another possibility is to place the camera so that the sunlight creates a cross light on the subject. (See Figure 2.)

Film can be shot in a normal lighting balance or in a high-key or a low-key lighting ratio. Complete control over lighting is one means of controlling the color negative image. However, no matter what image the lighting creates, that image can be further enhanced with camera filters and laboratory techniques. When you know what the post production requirements are in regard to optical work, such as a matte shot, you may have to alter your lighting balances somewhat. This topic is discussed in greater detail in Chapter 7.

Lighting is mentioned here because it is an important control factor—so important, in fact, that some directors and directors of photography intentionally schedule the filming of certain exterior scenes early in the morning or late in the afternoon when the sun is lower and a beautiful cross light is created. The warm color of early or late sunlight can enhance the mood of a scene. What happens during the other hours of the day? The producer need not panic! Other scenes are filmed: scenes that can be matched to early or late sunlight, different scenes that don't have to match, or perhaps interior scenes.

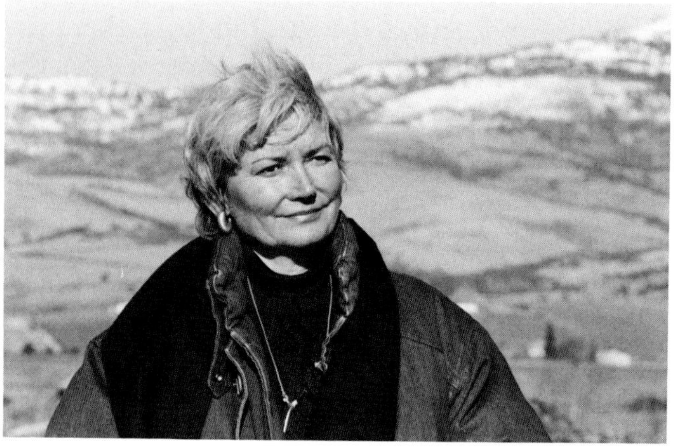

A

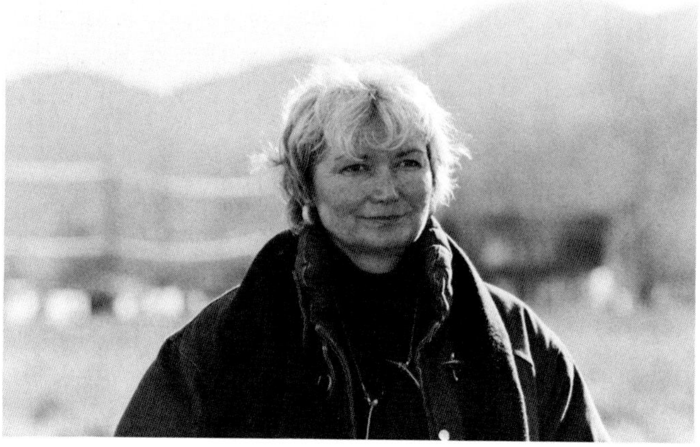

B

Figure 2 Using the sun as a flat light gives the best color rendering, as shown in Figure **2A**, but the scene lacks depth and the direct sun may cause the actor to squint. For Figure **2B**, an almost direct back light, with no reflector or other light source was used. The exposure was adjusted only for the face. Note: Using the sun as a back light rims the actor with light, and opening the lens to compensate for back lighting makes the background more out of focus, creating a greater three-dimensional effect. Figure **2C** was filmed in a cross light: a dramatic use of the sun's position. In some instances you'll want to use some fill light for the shadow areas to reduce the contrast. In this case, an H&H double fog #1/2 was used to reduce the contrast. (See Chapter 4.)

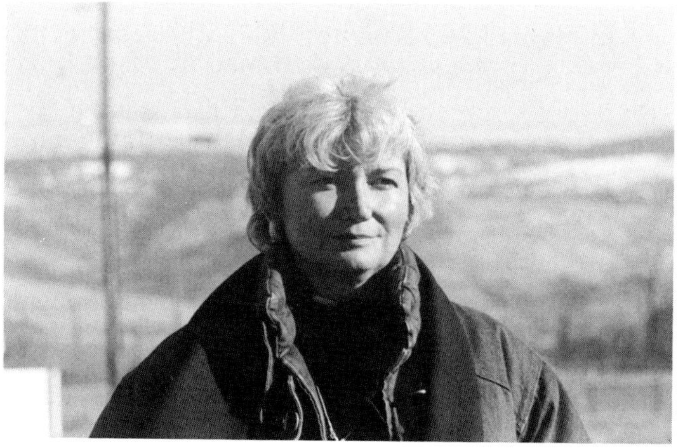

C

OTHER FACTORS AFFECTING THE IMAGE

The use of soft light, whether indoors or out, creates slightly muted colors. A point source of light, such as the spectral source of sunlight, gives a more saturated color response than that of a broad reflected source: the difference between a cloudless day and a bright overcast. Another means of softening the colors of the scene without resorting to a camera or lab technique is to add something to the atmosphere that not only spreads the light but, due to the nature of the effect, adds a bit of white light to the colors and thereby mutes, or slightly desaturates, them. This softening of light and color can be achieved by generating smoke or fog at a low level so that it's not actually visible, but softens the look of the scene. Several directors of photography use a low level of artificial fog or special cinematic smoke in many of their interior sets. The colors are very slightly muted, and the pictorial effect is lovely. By filming through a greater density of fog or smoke, an additional illusion of depth is created because the background is more veiled.

Haze, smoke, fog, and moisture suspended in the atmosphere and maintained for the period of filming have a different effect on the lens than that of filters. With natural or created haze, the objects closer to the lens show little or no effect; however, as the subject's distance from the camera increases, the effect becomes increasingly stronger, thus muting the colors and softening the sharpness of subjects farther from the lens. If you don't have the necessary equipment or the time to create a haze, you can create a similar effect with camera filters.

Another way to alter the normal color negative response is by using colored gels on the lights—whether key lights, back lights, or something in between. There may be times when there are no lights, no fog machines, and no overhead silks or scrims and the cinematographer's only recourse is the knowledge of what a certain filter or film lab can do to help give an average-looking image a distinctive look.

With every kind of film, the sooner the director of photography starts to work, the greater the chances of a coherent, integrated production. Cinematographers look at all the physical elements very differently from the others involved in the production. By visualizing the final release print, the cinematographer makes many decisions designed to tie together all the individual elements into one unified whole. It's really astounding how many decisions must be made before running even one frame of film through the camera!

Following is a review of the controls discussed thus far, with additional ideas for a preproduction check list. The cinematographer's job is to evaluate all the options and learn which ones are possible from a production standpoint.

1. Note in your script all of your ideas for special effects, moods, looks, transitions, and various script and character requirements. Transcribe the notes to a separate list that can serve as the basis for your discussions with others.
2. Once you and the director have determined just how you would like the film to look, confer with the production designer, costume designer, and set decorator on a color scheme to support and be supported by the photography.
3. If it's agreed to have day exteriors filmed on overcast days, meet with the production manager and the assistant director, who set up the shooting schedule. Whatever effect you seek—early morning, late afternoon, bright mid-day sun, dawn, dusk, whatever—make them aware of the visual style and your shooting needs. Production department personnel are always glad to help when they know your requirements in advance.
4. List the equipment you will need to carry out the overall plan:

    camera
    lenses
    raw stock
    camera filters
    lighting equipment
    grip equipment
    special camera gear
    special effects equipment

5. Last, choose the film lab, if you have a choice.

All of these elements affect the look of your film, and they represent the first steps to be taken toward image control.

Chapter 3
Image Control at the Camera

Once the principal photography of a film has begun, the director of photography is asked to make a great number of decisions. Many of the questions asked by various members of the crew are relatively easy to answer. They usually refer to the placement of the camera, lights, props, or special effects. The more difficult questions are the ones cinematographers ask themselves, and these mainly refer to the look they are trying to create for the film. These questions concern the choice of lens; camera movement; what filters to use, if any; the proper exposure; and the lab instructions. This chapter offers some ideas that may help new directors of photography make some of these daily decisions.

RAW STOCK

Before thinking about camera or lab techniques, you need to select the film stock for the production. The production manager has to know your decision so that he or she can place the order with the supplier. You might think that the choice of raw stock isn't related to controlling the image, but it definitely is a major consideration.

Due to the exigencies of commercial competition, cinematographers have an additional opportunity to govern the look of their films. They have at their disposal the products of different color negative manufacturers: Eastman Kodak, Fuji Film, and Agfa Gevaert.

Eastman makes a film that very nearly replicates the colors of nature. However, as mentioned in Chapter 1, when seen in a dark theater with the screen framed in black, the color seems too intense to some cinematographers. When the first Japanese color film, *Gate of Hell,* was released in the United States in 1955, the visuals of the film inspired outstandingly fine film reviews, and special mention was made of the color. Although shot on Eastman stock, special processing by the Far East Lab in Tokyo produced muted colors. When seen in a dark theater, the film seemed natural and beautiful. Fuji now makes a color negative that offers a less saturated, more muted color image that, in a darkened theater, offers the same natural look.

For many years, cinematographers had the choice of using either Eastman #5247 with a tungsten exposure index (EI) of 125 or Fuji #8511 also with a tungsten EI of 125. If a more saturated, brighter color image was desired, the Eastman product was used with normal processing. If softer, more muted, less saturated color was preferred, the Fuji negative was chosen.

Now we have an even greater choice. In addition to the emulsions just described, Agfa, Eastman, and Fuji offer "super-speed" color negative emulsions:

Eastman
#5296—35mm—tungsten—EI 500
#7292—16mm—tungsten—EI 320

Fuji
#F-250—35mm—tungsten—EI 250
#8514—35mm—tungsten—EI 500
#8524—16mm—tungsten—EI 500

Agfa Gevaert
#XT-125—35mm and 16mm—tungsten—EI 125
#XT-320—35mm and 16mm—tungsten—EI 320

Eastman #5296 has an EI of 500; however, many cinematographers increase it to 640 or even 800, which still prints in the middle of the printing scale of many film labs. It's important to make a definitive lab test (see Chapter 8) to ascertain that the way you use your exposure meter, the degree of shutter angle of your camera, and most important, the way the film lab processes your color negative, are all accurate, individually and in combination. The lab test will tell you the proper exposure index for the raw stock you plan to use, based on how you plan to expose it for the look you want.

When Fuji first came out with its high-speed color negative, it still had that wonderful look of less saturated color; now, however, it's almost the same as the Eastman high-speed product as far as color is concerned. Fuji adjusted the emulsion, losing some of the more muted colors, in response to some cinematographers who said that they couldn't get enough black when filming night exteriors with available or limited light. Naturally, as colors become muted, the black becomes a little less black and appears as a very dark gray. I never thought that this criticism was justification enough to change an excellent emulsion. Because Eastman and Fuji emulsions were quite compatible in all other respects, cinematographers could have filmed with Fuji for the muted color response and, when a scene required rich blacks, switch to Eastman. I've put this technique to good practice, using Fuji #8511 for day exteriors when I wanted softer color and switching to Eastman #5294 (prior to the release of #5296) for night exteriors, in *The Ultimate Warrior* and *The Car.*

In addition to these emulsions, all manufacturers have newer emulsions in both 35mm and 16mm that have finer grain structure, more stable dyes, and a range of exposure indexes. They also have emulsions designed for daylight color balance.

Fuji recently introduced five technologically advanced emulsions, which are listed here. As of this writing, I have little information about them beyond their nomenclature. I did shoot a lab test with Fuji #8514 that turned out very well. (You can see the results in Color Plates 23 and 24.) These are the F-series, 35mm emulsions:

Fujicolor Negative Film
F-64, #8510 (3200°K) EI 64
F-64D, #8520 (5600°K, daylight) EI 64
F-125, #8530 (3200°K) EI 125
F-250, #8550 (3200°K) EI 250
F-500, #8514 (3200°K) EI 500

All of these films are designed to be technically superior to the earlier emulsions. Additional information can be obtained from

Fuji Photo Film U.S.A., Inc.
1211 East Artesia Blvd.
Carson, CA 90746
Tel.: (714) 372-4282

Eastman has introduced four new extended-range (EXR) emulsions:

- #5296 (35mm) has a recommended EI of 500 in 3200°K tungsten light. Eastman claims that it's their sharpest, finest-grained, and fastest high-speed color negative, a true EI 500 film, which may mean that under certain conditions and tests it could be exposed at even higher exposure indexes.
- #7248 (16mm) has a recommended EI of 100 in 3200ºK tungsten light. This emulsion replaces Eastman #7291 color negative.
- #5245 (35mm) and #7245 (16mm) both have EIs of 50 and are balanced for exposure in daylight. Eastman claims that these are the best color negative films ever made for clarity, ultrafine grain, and sharpness.

These new emulsions are manufactured with Eastman's proprietary process: T-grain silver halide crystal structure, advanced color coupler technology, and their innovative Keykode numbers. The T-grain emulsion uses a flatter version of the silver halide crystal, which results in a sharper image with ultrafine grain. The EXR emulsions incorporate new generations of couplers and improved absorbing dyes. Image sharpness is improved by chemically inhibiting color dyes from migrating beyond the edges of the photographic image. The darker edges make the entire object appear to be more sharply defined. Keykode numbers on the new EXR films use a new edge-numbering system that is made up of both digital and operator-readable characters on the negative. The Keykode consists of a bar code that allows numbers to be read automatically by a scanner used during film-to-video transfers, negative matching, color timing, and printing.

In addition to these advantages of the new EXR emulsions, they also offer greater underexposure latitude. To quote Eastman,

> Ideally, cinematographers would expose every negative using the recommended exposure index. This would yield the best image quality for a particular negative in a given situation. However, for practical and artistic purposes, cinematographers will often intentionally underexpose film. EXR films were, therefore, designed with a greater underexposure latitude to allow the cinematographer to underexpose a scene and still capture exceptional detail at both ends of the exposure scale. EXR films allow you to show detail in both highlights and shadows while retaining sharp, fine-grained images with lush blacks and naturally saturated colors while avoiding build-up of grain, particularly in the shadow areas.

These Eastman and Fuji products are exceptionally fine new emulsions. I recommend that you make your own tests to determine an exposure index for your specific use of the film. As with any new emulsion, a comparison test with an emulsion that you've used before will show the advantages of the new product. For further information, contact Eastman Kodak or Fuji.

In *The Car*, the last 20 minutes of the story featured a chase through the countryside that began at the very first light of dawn and ended just as the sun broke over the horizon. This was one-fifth of the entire film, and we had a big problem! There was no way we could shoot only during dawn or dusk hours; we had to shoot all day long for about two weeks. I devised a plan using a combination of lab techniques, filters, and exposure changes. I used Fuji for the first part of the chase and then switched to Eastman for the last part of the sequence. The colors of the early morning naturally appear more muted than later in the day, so Fuji met this requirement. For the stronger, more saturated look just before sunrise, Eastman stock was used. (See "Image Control Schedules" in Chapter 10.)

Film manufacturers will usually offer a 400-foot roll of any emulsion for testing, in the hope that they will get the raw stock order for the film. Emulsion tests are another tool that directors of photography can use to control the image.

EXPOSURE

As simple a tool as exposure can drastically affect the image you put on the screen. Film manufacturers assign to their products an exposure index that indicates the film's sensitivity to light. Of the five systems currently used, the first two relate to the U.S.A.:

EI (exposure index)
ASA (American Standards Association)
ISO (International Standards Organization)
EV (exposure value)
DIN (Deutchen International
Normanausschuss)

According to the manufacturers' own recommendations, these sensitivity ratings should be used only as guides. Because of variables in lab processing, shutter angles, exposure meters themselves, and the ways in which they are used, the sensitivity ratings cannot be construed as an absolute certainty. Sensitivity ratings offer a guide for setting an exposure meter to produce a normal-density negative with recommended film processing. This exposure guide is based on creating a normal-density negative relative to flesh tone. Lab tests enable cinematographers to decide whether a different speed rating would be better for the density of a particular negative.

Laboratory printing equipment has a light range of 1 to 50. Exposed film that produces a negative that the lab considers to be "of normal density" prints in the "normal printer light range" of that lab. The overall printing range can vary slightly from lab to lab depending on their equipment and their processing, but lab technicians generally like to see a negative print between 30 and 37 on their printer lights. The number indicates the intensity of the light of the printing machine. The higher the number, the greater the intensity of light needed to correctly print a dense negative.

Actually, it's never just one number because the color negative is printed by an additive light process with sources of magenta, cyan, and yellow light that, when blended together, create white light. Each color has its own printer light number, so a color negative could print on printer lights of 30 magenta, 33 cyan, and 28 yellow. This is usually written on the lab report sheet as 30–33–28. The fact that the three colors almost never have the same printer light number is due to slight differences between emulsion batches of the negative or print raw stock, slight differences in the filters used in the printer, or possible differences in the lab processing. The lab's normal printer lights are determined by the lab's own testing of different negative and print stock emulsions to produce an image whose densitometer readings indicate that they coincide with set standards.

As just mentioned, printing equipment has an overall range of light densities from 1 to 50, and as far as the lab is concerned, 30 falls in the middle. In the days of black-and-white cinematography, 25 was the midrange number, but color emulsions shifted the midpoint. The new midpoint, 30, gives the lab leeway to go up or down, depending on the density of your negative or the effect you want. There are eight printer lights for each stop of exposure. The more dense the negative, the higher the printer light number required for a normal-looking print. Conversely, a less dense negative will print on lower numbers. An overexposure of one stop requires eight higher printer lights for a normal-looking print; an underexposed negative of one-half stop requires four lower printer lights. (See Chapter 8.)

Something happens when the numbers get too low, however, which indicates a greatly underexposed negative. When the lab tries to produce a normal print, the colors are not as saturated, and the blacks are a bit smoky or slightly greenish or grayish. However, if the director of photography was intentionally trying to create an effect such as a moody night exterior or a low-key interior and asked the lab to print the underexposed scene on higher printer lights, the result would be different: there would be a true black, the colors would be more saturated, and the mood of the scene would be low key.

Besides creating low-key scenes, underexposing is the key to producing a good day-for-night effect. Positioning the camera correctly in relation to the sun and using filters to reduce the brightness of the sky, you can underexpose the scene by one and a half to two stops and instruct the lab to print the film in the normal range. The result is a very successful night effect with the convenience and cost effectiveness of filming in daylight. (See "Creating a Day-for-Night Effect" in Chapter 5; see also Color Plate 2.)

Remember, good black density is a product of the printer lights, not a result of exposure! By increasing the printer lights, a good black level can be achieved whether the negative was underexposed or overexposed. By purposely underexposing the negative and instructing the lab to print for a particular effect, you alter the normal response of the color film. If you were to overexpose the negative intentionally and ask the lab to print it on their normal lights, you'd see a print that was bright, the colors a bit washed out, and perhaps no black blacks. This would be fine if you wanted a very high-key effect or a special look. If you wanted a normal look from the overexposed negative, the lab would use higher printer lights. Depending on the degree of overexposure, you can create a more saturated color. Even if you accidentally overexpose by as many as two stops, you can still try printing for a normal look. There may be some color shifts, however, especially in the sky and the pastel range .

In an otherwise normally exposed scene, if a few frames are purposely overexposed by momentarily flashing them with an extremely intense light, such as an arc, the print will show a lightning effect. If a long-peak flashbulb is set off, the high intensity of the light will create a flash effect in just a frame or two. The normal exposure is altered, and you've created an effect.

The *American Cinematographer Manual, Sixth Edition,* edited by Fred H. Detmers, (Hollywood: The A.S.C. Press, 1986, p. 66), states the following:

> Generally, use of ASA lower than manufacturers' recommendation (over-exposure) will produce finer grain, higher color saturation, and a slight increase in sharpness at the expense of loss of highlight detail and flattening of whites. Use of a higher ASA than recommended (under-exposure) will show more grain, lower color saturation, loss of sharpness and loss of shadow detail.

Relative position on a laboratory printer scale is also a factor. It depends on where you instruct the lab to print for the desired effect. By shifting the standards of normal exposure, the cinematographer has another method with which to alter the color negative response. (See also Chapters 4 and 6.)

A subtle technique for controlling the contrast of a scene is to vary the F/stop. If the cinematographer is aiming for a normal density negative, the contrast can be affected by changing the F/stop or T/stop. In exterior scenes, many cinematographers make it a point not to stop down past F/5.6 even though the lens can be shut down to much smaller stops. They know that with small lens openings, F/11 to F/22, the contrast of the scene is greater than when filming at more open stops. This is admittedly a subtle effect, but it can affect the visuals of a scene.

Proper filtration and/or the use of neutral density filters keeps the exposure normal when you want to shoot in bright sunshine with the lens wide open. However, if you are using a zoom or telephoto lens, if you want as much depth of field as possible, or if you want to help the camera assistant with a difficult follow-focus shot, you will stop down as much as possible. With long focal length lenses, the slight increase in contrast resulting from a small lens stop might be advantageous because some long lenses inherently have slightly less contrast.

To sum up, a scene filmed at F/16 or F/22 will appear to have more contrast than the same scene filmed at F/4 or F/5.6. Make your own tests to verify this. Exposure is thus another means of creating a special look and controlling the image.

LENSES

Image control is also gained through the judicious use of lenses of different focal lengths. Not too many years ago, most professional 35mm motion picture cameras (Mitchell and Bell & Howell) had a lens turret on the front of the camera on which were mounted four different lenses. The turret could be rotated, enabling the camera operator to switch easily from one focal length to another. Those were the days before zoom lenses and single lens reflex cameras were widely used. The lens turret enabled the cinematographer to quickly change from a wide angle lens, used for the establishing shot, to the longer focal lengths used for closer shots. Not only did the directors of photography rotate the lens turret, they also moved the camera to obtain a slightly different background for the closer shot. Times have changed! Today, the speed of production too often takes precedence over the quality of the photography. The zoom lens has not only replaced the lens turret but all too often is used in place of a dolly shot. A scene filmed with a zoom lens, zooming from wide to close-up, appears to move the subject toward the camera; when a camera dollies in, the camera moves to the subject. The difference is very apparent on the screen: a dolly shot correctly changes the perspective; a zoom shot does not.

Cinematographers are often less likely to change lenses than to move the camera closer to the subject, even though changing a lens can require less than one minute. If the size of the subject remains the same, what happens to it if the lens is switched from a 35mm lens (used for a medium shot) to a 100mm lens (for a close-up), instead of moving the camera in closer with the 35mm? These are the differences between the two lenses in this instance:

- The 35mm lens has a relatively wide angle of

view, which means it will show quite a bit of background behind the actor's face. This is fine if the background information is important to the scene. If not, it can be a distraction.

- The 100mm lens has a narrower angle of view, which means it shows less background behind the actor's face. Background information or action is far less noticeable, reducing distractions. Also, you don't have to move the camera very much.
- The depth of field of the two lenses is drastically different. A background filmed with a 35mm lens is much sharper than with a 100mm lens. Not only does the 100mm lens show less background, it also enhances the impression of three dimensions by making the in-focus subject stand out against the out-of-focus background. (See Color Plate 3.)

When selecting a lens for a close-up, a prime consideration is what it will do to facial features. Unless you want to have a weird look, as in a horror movie, a wide-angle lens will distort a face unpleasantly when used too close. In the old Academy aperture format, a 50mm lens was considered most similar to the perspective of the human eye and therefore the most normal-looking; however, in today's wide-screen formats where faces appear much larger, the longer lenses are more flattering for close-ups.

When a zoom lens is used as the prime lens (that is, when it's used at various fixed focal lengths for different shots), some production time may be saved, but thought must be given to the selection of the focal length that is best for the close-up. A close-up may sometimes require a fair amount of background; in this case, stay wide on the zoom, and move the camera in. On the other hand, if the most flattering image of the actor is the foremost consideration, the longer focal length does a better job.

Most experienced cinematographers shoot definitive tests of all the lenses they plan to use on a new production, not only to see if the lenses are sharp and properly calibrated for focus but also to check the inherent contrast and to recognize any mismatch in color rendition. (See Chapter 8.) If a lens does not match in terms of color fidelity, the image can be corrected during the final timing of the film. To have scenes match when seeing dailies, you can install a color correction filter on the back element of the lens in question, thereby bringing it into balance with the other lenses.

The selection of a lens is another powerful tool for cinematographers to use for image control. All of the techniques discussed thus far involve choices you can make without adding any lens accessories, such as filters or diffusion, or asking the lab for any special laboratory processes.

Chapter 4
Image Control with Filters

Now we come to the means of color negative control that offers cinematographers the widest range of choices: filters. Once the lighting has been set, the lens selected, and the camera positioned, a camera filter—which modifies the light that reaches the color negative—can enhance the image.

It didn't take directors of photography long to realize that they had to have some means of controlling the light that passed through their lenses. Some of the problems they faced in the days of black-and-white film were beautiful blue skys that recorded as white, romantic actresses who looked too harsh in extreme close-ups, and green trees that looked darker than normal. Many early films were Westerns filmed mostly outdoors in plenty of light. Cinematographers needed a way to turn day into night for many scenes, and as films became more sophisticated vis-à-vis locations, additional effects were needed to improve the image. Cinematographers began using filters on their cameras.

A filter is not only a piece of tinted glass placed in front of a lens. A filter is any material that alters the quality of light passing through the lens to the sensitive emulsion. A filter can be placed within the lens elements, as in the case of a lens with a long focal length, or in a gelatin filter slot in front of the camera film aperture. It can be made of glass that has been impregnated with a color dye, a gelatinous material laminated between two pieces of optical glass, acrylic, fine net fabric, or semirigid gelatin. A filter can be made of any material that works for the cinematographer. All filters are not necessarily colored. Neutral density (ND) filters affect all colors equally, and slightly milky filters are used to create fog effects or to reduce the visual contrast of a scene. Other filters just affect the resolution of the lens and diffuse the image.

BLACK-AND-WHITE FILTRATION

The use of filters is by no means a new technique, nor is it limited to color film. Many directors of photography who began their careers in the days when only black-and-white emulsions were available were very well versed in the use of filters to help create various visual effects. Filters played a more important role in getting a normal black-and-white scene to look "right" than in creating a special effect.

The early film emulsions were orthochromatic; that is, they were not sensitive to red. You may remember seeing black-and-white films in which the heroine's lips looked black. In the early days of filmmaking, the interior sets were lit with carbon arc lights or mercury vapor lamps, both of which had a high content of blue light to which the early emulsions were most sensitive. The use of colored filters with orthochromatic film was very rare. However, filters were used for diffusion and special effects, such as fog.

Later, black-and-white panchromatic negative stocks were introduced; these were most sensitive to blue light, sensitive to red, and least sensitive to green. Due to the panchromatic emulsion's high sensitivity to blue, a blue sky would usually reproduce as white on the print, instead of a good medium gray. To correct this false rendering, a yellow filter was used. It held back some of the blue light of the sky, and when printed, the sky was a light gray. If a stronger effect was desired—that is, a darker sky for a more dramatic look—an orange or red-orange filter was used; however, the cinematographer had to exercise caution if there were light flesh tones in the same scene because they became lighter as the heavier filters were used. Although the orange filter held back the blue light of the sky, it permitted the red color of flesh tone to record more fully on the negative, thereby making the flesh tone much lighter on the print.

For scenic shots with a predominance of green foliage, a yellow-green to deep green filter was used to transmit more green to the image and at the same time hold back part of the blue and red light, thus allowing trees and foliage to look more natural. If there was an actor in the foreground, caution again was exercised because these filters made lips and skin tones look darker than normal.

All of these filters were used to record colors in various shades of gray that represented their corresponding colors and therefore looked very natural. Special effects were created by using deeper colored filters to make clouds stand out dramatically or to blacken the sky when doing day-for-night effects. The filters used in black-and-white photography did more than enhance the reproduction of colors; they were used for diffusion and fog as well. In those days, a fog filter was used just as its name implies: to create a fog effect. Today, when filming in color, filters similar to the old fogs are used to control contrast and to mute color.

COLOR AND THE NEED FOR NEW FILTERS

With the advent of color motion pictures, directors of photography and color technicians were primarily concerned with creating a sharp image with saturated color. Many of those early films attest to this. The first true color features were made by the Technicolor process, and oddly enough, black-and-white film was used in the famous three-strip Technicolor camera. This ingenious camera ran three strips of black-and-white film at the same time; each recorded a primary color: red, green, and blue. Through the process of imbibition printing, dyes were transferred from three special dye-absorbing prints to the final print emulsion, or matrix. By superimposing one layer of dye on another, the final result was a full-color print. The process was similar to today's techniques for printing color illustrations. The three-strip Technicolor camera was very cumbersome for some location work, so Technicolor introduced mono-pack color film. This daylight-balanced color emulsion was capable of recording a full-color positive image on one piece of raw stock that worked in normal cameras. The positive color original was used to make three black-and-white color separation negatives that were printed by the standard Technicolor process.

Early documentary and travel films that were shot in color used 16mm color reversal film. There was no color negative, and when making multiple prints and optical effects, controlling contrast and color reproduction was always a problem. The labs could do only limited color correction, so cinematographers had to get the best original color possible on the color reversal film. Color-compensating filters were available then (and still are available) to balance the film to the available light source. These filters were made in various densities of magenta, cyan, yellow, blue, and red. At that time, few cinematographers thought of creating color responses other than those that were most realistic and natural. They were delighted to be able to create a normal color image in abnormal light conditions. When filming in bluish overcast, matching an early or late sunlight to midday, or shooting interiors with artificial light, they were able to match the color emulsion to the prevailing light condition with the use of filters.

As the industry developed and the film manufacturers made 35mm and 16mm color negatives, other changes were taking place. The television industry advanced from black-and-white to full-color transmission. This created an exciting experimental atmosphere for cinematographers, who used television commercials as their research tools. Commercials, which served as subsidized 60-second mini-films, occasioned the development of new techniques as a result of the growing demands from advertising agencies and their clients. The commercial is the richest, most fertile training ground for film innovation! Eventually, a new breed of cinematographer moved from television commercials to feature films. Many of today's finest directors of photography got their first experience filming television commercials. The changing styles that have developed in theatrical film production and in television have caused cinematographers to call upon filter manufacturers to create completely new types of filters.

For a long time in the United States there were four prominent filter manufacturers: Harrison & Harrison and Pancro-Mirror in California, Tiffen and Eastman Kodak in New York. Newer filter companies have been established in response to the greater demand for filters; they're listed in the Appendix. All of these companies make similar filters, yet each has its own specialties, and even with the same type of filter, there are certain differences. The main distinctions between the four manufacturers are these: Eastman Kodak is primarily concerned with color-conversion filters, that is, filters made of gelatin or glass used for scientific and many other purposes and filters made specifically for still photographers. Pancro-Mirror doesn't make any colored filters; they specialize in neutral density filters and diffusion glass of all types and sizes. Harrison & Harrison (H&H) and Tiffen, the two companies that manufacture the majority of the filters used by professional cinematographers today, both have very large selections and will produce to specification any filter requested by a cinematographer.

Filming for television offers a rich opportunity for using filters. The tonal range that television cameras and receivers can handle is limited compared to film. Film can handle a tonal range of about 140:1; television, only about 40:1. The difference between the darkest area and the brightest area represents the contrast. Because television can handle only about one-third the tonal range of film, some adjustments must be made when filming for television. Directors of photography can compensate for the restrictive tonal range by reducing the contrast of their TV films by controlling the lighting or by using filters. Eastman Kodak recently produced a special print stock (Eastman #5380 for 35mm and #7380 for 16mm) that provides lower-than-normal contrast; films not originally made for television can be printed on this stock, which reduces the contrast range to a more acceptable level for video transmission. The electronic industry advances and improves quickly. It may eventually be able to match the contrast range of film.

Contemporary filmmaking, whether for theaters or television, usually includes the use of a large variety of filters and effects. Many directors of photography carry a wide assortment of filters with them so they can maintain a consistent look from scene to scene under variable light conditions, enhance their lighting, or create a special look.

There are several ways to mount a filter on a camera lens, the most common being to use a matte box, which is a combination filter holder and lens sunshade. The matte box attaches to the camera by rods that adjust to lenses of different focal lengths, and will often accomodate three separate filters, round, square, or rectangular, in a variety of sizes. In some matte boxes the filter holders slide left and right, up and down, or rotate, in order to adjust a graduated filter to the horizon line. (See Figure 3.)

Round filters can also be screwed directly onto a lens using the lens mount threads. Or, a light weight filter holder and sunshade can be attached to the outside of the lens mount with special adapters. Filter holders are usually made in combination with a lens sunshade to prevent stray light from striking the filter and possibly spoiling the scene.

Filters can also be made of gelatin sheets that are cut to fit special holders that slide into a filter slot between the lens and the film plane. Many cinematographers prefer this method for standard filters because it eliminates any glass in front of the lens. When very long focal length lenses are used, a filter slot for gelatins is provided within the lens barrel itself because a glass filter on front will spoil the resolution of the lens. Sometimes a gelatin filter holder is attached to the rear lens element.

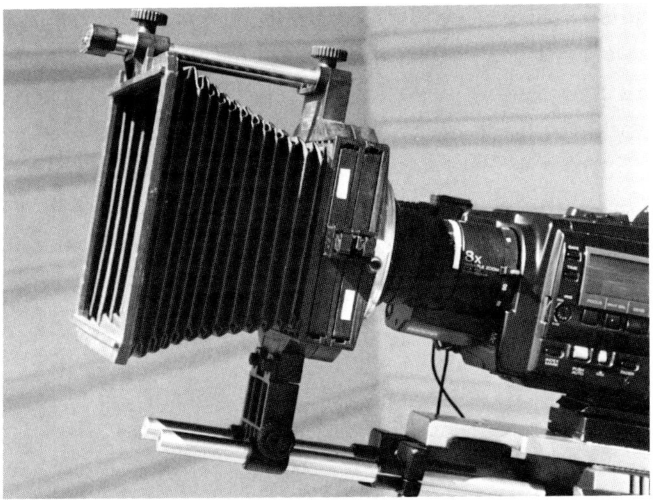

Figure 3 A typical matte box, fiilter holder (2 stages) and adjustable sunshade bellows.

FOG FILTERS

In an effort to reduce film contrast in color video images, camerapersons began using fog filters. This technique eventually found its way into feature films when cinematographers sought more muted colors. A fog filter does to contrast exactly what real fog does: it lends a slight milkiness to all the colors, which reduces the color saturation and thereby reduces the visual contrast. Fog filters are made in a variety of densities, from very slight to very heavy fog. It didn't take feature filmmakers long to realize that they could use this effect to create softer colors for certain types of scenes and to reduce the contrast for video transmission. (See Color Plate 4.)

Regular Fog Filters

Regular fog filters are made up of light-deflecting particles suspended in a gelatinous medium, which is then laminated between two pieces of optically clear glass in a patented process. These particles are able to scatter light almost in proportion to the brightness of the objects within the scene. The scattering of light gives the subject a bit of white light, which slightly mutes the color, depending on the strength of the filter selected. This is exactly what fog does in real life: the moisture in the air scatters light and veils what we see in a soft white haze. In real fog, a very bright object, like a street lamp, has a halo around it because the light from the lamp is more scattered than the surrounding light because of its intense brightness. Similarly, the regular fog filter creates a halo around any bright object in the scene, be it a lamp, window, white building, or even a white dress or shirt. Just as real fog diffuses our view of a landscape, the regular fog filter slightly softens the resolution, or sharpness, of the lens.

Cinematographers began to incorporate these mellow qualities in their work, thereby lending a different look to their films. Soon, due to post-production problems with matching opticals, the optical houses objected to the halos around bright objects and the slight diffusion caused by the regular fog filters. These filters also added a fringe of scattered light on the edge of any dark area with a bright background, for example, a tree silhouetted against a hazy sky or a person in dark clothes against a bright building. But cinematographers still liked the muting effect of the regular fog, so they prevailed on filter manufacturers to produce a filter that would still desaturate both light and dark colors, but would also eliminate the halo and the diffusion effects. Filter manufacturers soon complied with their request and created the double fog filter. (See Color Plate 5.)

Double Fog Filters

The double fog filter was made in a variety of densities and could be used exactly like the regular fog but without its objectionable characteristics. The double fog and the regular fog filters affect both the shadow and the highlight areas in direct proportion to the strength of the filter selected. The lower the number of the filter, the weaker its effect. Filter sets consist of #1 through #5; however, filter manufacturers have often been asked to produce fractional fog filters, such as ⅛, ¼, ½, and even 3⁄16.

To desaturate colors only slightly, ¼ through #1

or #2 regular or double fog filters can be used. For stronger desaturation, heavier filters (higher numbers) are used. My favorite filter for *Diary of a Mad Housewife* was a #3 regular fog. All of the regular fogs soften the resolution of the lens very slightly compared to the double fogs. For this reason, I generally use a ½ double fog, which slightly reduces the contrast while preserving the sharpness, when filming for television. Fog filters can also be used in combination with other filters to create more complex effects.

LOW-CONTRAST FILTERS

As the film and television industries expanded, new emulsions were created to expedite the processing of greater volumes of film. A new developing process was concurrently devised that used a warm bath and required a much shorter developing time. These two innovations made it possible to process film less expensively, offsetting escalating labor costs, and to process film faster, as the TV industry demanded. However, the new emulsions and the new developing process changed some characteristics of the color negative. The color negative and color print stocks—although technically improved as far as resolution, color saturation, and storage capability were concerned—were also quite a bit more contrasty. Many cinematographers, unhappy with this characteristic of the new color stocks, again called on filter manufacturers for a solution. They responded with a new filter that affected the darker shadow areas but not the lighter highlight areas. The low-contrast filter reduced the contrast by reducing the tonal range of the scene and gave cinematographers a new tool.

To take full advantage of the low-contrast filter (low-con), it's important to remember that it affects only the shadow and dark areas, not the highlight and bright areas; thus, it reduces the tonal range by making the shadow areas slightly lighter. Exposure also plays a part here.

Imagine a white house with a background of heavy foliage and trees. A low-con filter, plus underexposing one-half stop, will help control the contrast of the scene. This is how it works. The low-con adds light to the dark area (trees and foliage) without affecting the bright area (the house). The house, unaffected by the filter, is less bright when underexposed one-half stop, and because the low con adds brightness to the foliage, the underexposure brings the foliage density back to where it would have been without the filter. The low-contrast filter, when coupled with slight underexposure, reduces the tonal range, or contrast, of the scene; it is not intended to desaturate the color in this case.

However, if the goal is color desaturation, one-half stop overexposure allows the low-con to act more fully on the darker colors, adding a bit of white light and thereby muting them. The exact result depends on the strength of the filter selected. Lighter pastel colors are affected little by low-contrast filters. Exposure and printing play important roles in the effectiveness of both fogs and low-cons. (See Color Plate 6.)

Both H&H and Tiffen manufacture a series of low-contrast filters, the uses of which are limited only by the ideas of the cinematographers. Although designed for the same purposes, there are some differences between the filters of these two manufacturers.

I use the low-cons from both manufacturers to use the differences to advantage. Both low-contrast filter sets have a normal range of #1 through #5, and both manufacturers will make, on special order, any density requested—lighter than #1 and heavier than #5, or any fractions in between. My set of H&H low-cons consists of ⅛, 3/16, ¼, ½, 1, 2, 2½, 3, 4; and 5. My Tiffen low-con series is from #1 through #5. I've found that the difference between these two series is twofold.

First, the H&H filters give a stronger low-contrast effect than the Tiffen filters. I find a Tiffen low con #4 or #5 to be equal in effect to an H&H low con #1½. The advantage is that different filters can be combined to create just the right degree of contrast reduction when one filter is not quite enough and the next strongest is too much. If you have

H&H #2 and #3 and you want a #2½, try an H&H #2 with a Tiffen #2.

The second difference is the ability of the Tiffen low-con to work better than the H&H in strong back light situations or where bright areas—such as sky, sand, snow, or a bright window—tend to make the foreground image a bit milky. In these instances, an overall flare appears in the H&H low-con that causes a milky look.

Through testing, you can evaluate both series of low-cons, and perhaps you will use them both, as I do. I find that both series are essential when working with multiple cameras and trying to match the contrasts in a situation where the cross light at camera A becomes the back light of camera B and the front light of camera C!

A word of caution: Special care must be taken to prevent direct light from striking any of these filters as it will result in an extremely washed-out image. Make sure that your assistant properly protects the filter in the matte box by adding an additional front shield or by setting up gobos in front of any lights that strike the filter. When using a matte box, be sure to keep any extraneous light from entering behind the filter around the lens. Panavision makes a soft rubber "donut" that fits around the lens and seats against the matte box to form a light seal. (See Color Plate 7.)

When planning a shot in which the source of light is included in the frame, the use of any fog or low-contrast filter is a major consideration. The light source will often reduce the contrast by itself, and if a low-con or fog is used, the light hitting the filter can make the scene appear very milky. Look at the scene through the lens to determine whether to use a particular filter. If matching color desaturation is the most important factor and a low-contrast filter will spoil the scene with the extreme milkiness created when a photographed light strikes it, you might try using a light white net as a matte box filter. The scene won't become as milky when light strikes the net, yet all of the darker colors will become somewhat desaturated. (See Chapter 5.) Another way to match color desaturation in such a case is to have the negative flashed. (See Chapter 6.)

MORE ABOUT FOGS AND LOW-CONS

Because all fog filters impart a bit of white light to all colors, light and dark, relative to the strength of the filter, the overall effect is of slightly increased exposure even when no exposure increase is made. By increasing the exposure one-half stop when using fog filters, the desaturation effect can be intensified. (See Color Plate 8.)

Interestingly, if you increase the exposure while using a medium-density low-contrast filter and the lab makes the print a bit darker than normal, you might find that the increased effect you wanted has been all but eliminated. Conversely, with the same low-contrast filter, if no increase in exposure is made and the lab makes the print slightly lighter than normal, you'll find an increased desaturation effect. It's therefore important to know in advance the ideal print density and to choose the filter strength and exposure levels accordingly. It's wise to inform the lab of your ideal print density and the effect you're seeking, as they can be of great help. Tests are the only true basis for establishing the "norm" for your effect.

Using a medium to heavy low-con often gives a slight "veiling" or muting effect to both colors and black when printed on normal printer lights; however, if the print is made three or four points darker than normal, the veiling will not be apparent, but the contrast will be reduced. This is called *printing through* the low-contrast effect.

Once you've decided on the degree of filtration for a particular sequence, you'll probably want to vary that effect to make the entire production more visually interesting. Filtering will vary with changing conditions of light, set colors, costume colors, different locations, weather variables, and, most importantly, the visual dramatics of the scene. Whatever effect you begin the sequence with, a wide assortment of filters will allow you to use various combinations and maintain a visual continuity throughout. You must judge the elements of each setup and select the correct degree of filtration to have the whole sequence match. This may seem difficult at first, but like many techniques, it becomes easier with experience.

During these days of tighter budgets and shorter shooting schedules, cinematographers do not have the luxury of not shooting when the weather changes from clear sun to bright overcast or to heavy overcast. Experienced directors of photography request the weather forecast from the local airport or weather station, but also keep an eye on the sky for signs of upcoming changes that may affect the shooting.

When light conditions are expected to change, you can be prepared. For example, if you start filming in clear sunshine, which creates contrasty dark shadows, an H&H low-con #2 can be of great help. If the weather changes to hazy sunshine, you'll notice that the shadows aren't as dark as before, so a better match will occur if you switch to an H&H low-con #1. If the weather continues to deteriorate, a still lighter filter will help match the next setup, possibly an H&H low-con #½ or a Tiffen low-con #2 or #3. If you continue filming and the weather worsens so that there's much less contrast in the scene, just shoot without any filter at all! Sometimes when you know that the weather is going to change, you can film an isolated shot or two until the clouds roll in, and then film the longer sequence entirely in overcast, which can look good.

Obviously, this method can't make an *exact* match from scene to scene because if you lose the sunshine, you also lose the shadows, but it does help keep the contrast in balance. If you have some arc or HMI lights, you can create your own shadows on your foreground subjects; or, in bright sunshine, use the lights to fill in the shadows somewhat. Lamps can always be used to help control contrast, but that's another subject.

Although there aren't any filters that *increase* contrast, there are steps you can take to accomplish that goal, if you must. Forced development slightly increases the contrast of the color negative, and making a dupe negative further increases the contrast. These laboratory controls are discussed in Chapter 6.

In review, fog filters work differently from low-contrast filters. The fogs affect both dark and light areas; the low-cons do not affect the highlight areas but do lighten the dark and shadow areas. Both are influenced by and work more strongly in relation to the brightness of the background—a bright sky, a light building, or light walls of a room—because the additional light from the background strikes all of the particles of the filter and then spills back to lighten the darker areas. For example, in a scene where two actors dressed in medium tones stand in front of a dark wood-paneled wall, you might use an H&H low-con #2 or a Tiffen low-con #3 or #4 to reduce the contrast of the scene for better television transmission. If the same actors in the same wardrobe stand in a kitchen set with very light walls and cabinets, you'll quickly see that these low-cons create too strong an effect on the actors. The amount of light within a scene is of utmost importance in relation to the effect created by a low-contrast, regular fog, or double fog filter! Aside from lighting, the visual contrast within the scene, created by different colors, also has a direct bearing on the choice of a fog or low-con. It's important to remember this as the camera is moved from setup to setup and backgrounds change.

The strength and effect of a regular fog, double fog, or low-contrast filter should be judged through the lens on the ground glass. Cinematographers must condition their eyes to recognize the change the filter makes, first by looking through the camera without the filter and then by having the assistant slide the filter in. If you concentrate on the dark areas, you'll soon be able to judge just how much lighter they can appear to be, thereby reducing the contrast and allowing you to match one scene to the next. Note that the filter effect is more exaggerated on the ground glass when the lens is wide open, and the effect is difficult to judge at all if the lens is stopped way down. As a general rule when filming outdoors in normal bright daylight, judge the filter effect with the lens stopped to about F/5.6. Indoors, when the exposure is about F/4 or F/4.5, judge the filter effect at F/3.2, but definitely not at F/1.4. If the ground glass image is too bright, it will create a false effect. By comparing what you saw through the camera to what you see on the screen, you can learn to what degree you should filter a scene.

When filming with low-cons or fogs and trying to match scene to scene, you can encounter a problem if you switch from a prime lens to a telephoto, or a very long lens, such as an 800mm or a 1200mm. If the lens is of good quality, it will produce a sharp image with excellent contrast. However, if you place a fog or low-contrast filter in front of a long lens, you can destroy the image sharpness beyond usability. These lenses cannot tolerate any glass in front of them. If a #85 filter is required, the lens usually has a filter slot in its lens barrel for a glass or gelatin filter, but anything in addition to that destroys the optics of the lens. How then can you reduce the contrast to match other scenes already photographed with a fog or low-contrast filter? There are two alternatives. You can flash a dupe negative in the lab, a technique described in Chapter 6. However, that's costly and time consuming and therefore may not be the best choice. The alternative is to use a net material in front of the lens. It will not affect the refraction of the lens, but it will reduce the contrast. Nets are discussed in "Diffusers" in Chapter 5.

In the above case, a net of white threads with coarse openings will very closely match the effect of a low-contrast filter. The wide mesh permits most of the usable light to pass unaffected, and the light that bounces off the white threads scatters into the colors and dark areas just enough to reduce the contrast without appreciably affecting the lens resolution. It's easy for cinematographers to carry an assortment of net materials on all film assignments and then use the one that most closely approximates the effect of a low-contrast or fog filter.

Recently, Tiffen introduced several series of filters that can also be used to reduce contrast: the ultra contrast filter, the pro mist filter, and the soft contrast filter. Each of these series has subtle effects. Ultra contrast filters lower the contrast by slightly brightening the shadow areas without creating flare or halo from light sources within the frame. Pro mist filters subtly reduce the contrast and slightly soften the image; the heavier filters slightly mute strong colors. Soft contrast filters reduce contrast within the scene without affecting the black areas, and they tend to soften intense colors. Tiffen offers technical data on all of their filters. (See the Appendix.)

Only by testing the various filters and using them on the job will you be able to master the techniques of changing the color negative response to reflect your own creative choices. In summary then, three types of filters are available—regular fog, double fog, and low-contrast—each of which affects the contrast and the colors within the scene with subtly different characteristics.

Real Fog Effect

Although fog filters are often used to desaturate colors, reduce contrast, and perhaps lend an air of fantasy to some scenes, you cannot completely ignore the primary reason for making the first fog filter: to create the look of real fog. Filters are rarely used today to create the semblance of real fog because special fog-generating equipment is available. However, certain points are worth covering.

When real fog occurs, there is no direct sunlight; therefore, an overcast day is necessary to create a natural-looking fog effect. Overcast also eliminates the problem of blue sky showing in the scene, which looks illogical. With real fog, which is a suspension of fine water particles in the air, the fog effect increases in proportion to the distance from the viewer; that is, the fog obscures more of the detail and color of distant objects than it does of foreground objects. A regular fog filter has the same effect on close objects as on distant ones; therefore, although it adds a soft overall haze, it doesn't mirror reality very effectively. When a scene requires real fog, the special effects department usually comes to the rescue with fog generators, which lay a fog-like layer of special vaporized liquid in the air. This is simple enough in a studio, but outdoors, if a wind develops, the artificial fog is blown away. Only on a virtually windless day can fogging equipment work properly outdoors.

Scenic Fog Filters

Fog filters are available that have variable densities within the same filter, from clear at the bottom to full density at top, with a blend anywhere in between. A graduated filter like this is usually made oversized (as all graduated filters should be) so it can be shifted within the matte box until the ideal position is found. With these filters, the foreground is clear, and the upper portion of the frame receives the greater fog effect. A normal overall density fog filter (such as a regular fog #2 or #3) can be combined with a graduated fog filter to create many different effects. If the framing happens to include a large object in the foreground, such as a house or tree, the filter can be turned 90° to keep that side clear, and the rest of the frame, reaching into the background, will have the increased fog effect. H&H recently made a new set of graduated fog filters called scenic fogs. (See Color Plate 9.)

Creating a true fog effect is usually the responsibility of the special effects department, and these experts know all the conditions that should prevail. However, an "untrue" fog effect can also work very successfully for certain scenes, and for these we thank the fog filters and scenic fogs that not only help us mute the colors and reduce contrast, but also create the appearance of something that isn't there at all!

COLOR FILM AND THE LIGHT SOURCE

The filters discussed thus far have no color themselves. It's now necessary to see why some colored filters are mandatory for achieving a normal color response from a color negative. After that, we will look at filters that change the color response for the additional purposes of enriching mood and creating emphasis.

A succinct explanation of the relationship between light sources and color films (both 16mm and 35mm) is found in the *American Cinematographer Handbook*, edited by Arthur C. Miller, A.S.C. and Walter Strenge, A.S.C. (Hollywood: A.S.C. Holding Corp., 1969, p. 448).

> Though virtually negligible in black and white, the color quality of the light source is an all important consideration in color photography. Color film does not always "see" colors as human beings see them. Humans think of color as being the same in both tungsten and daylight (actually a combination of sunlight and skylight). Human vision automatically compensates for the change from *yellow* tungsten illumination to *blue* daylight. Color films can make no such compensation. Color film produces color approximately as the eye sees it only when the illumination is the same as that for which the film is balanced. [Author's note: i.e., color negative balanced for 3200°K (tungsten) or for 5500°K (daylight).]

When using color negative, most color correction can be done later by the film lab. However, when using a color reversal film, such as Kodachrome or Ektachrome, it's more important to make the corrections at the time of shooting. Meters and charts are available to help the cameraperson decide which correction filter to use. In 35mm cinematography, most filming is done with a tungsten-balanced color negative, and when this film is used in daylight conditions, a #85 or #85B color-balancing filter is used. Any slight variations in the color of the light source, which would normally result in a screen image slightly off-color, are readily corrected later in the film lab instead of using a large variety of color-correction filters on the camera.

The Kelvin Scale

The color of illumination used for filming is rated in degrees Kelvin (°K) according to the Kelvin scale. A British scientist, Lord Kelvin, discovered that a black body such as a tungsten filament acquires different colors as it is heated. The metal first glows a dull red, then bright red, then orange, then yellow, then white, and finally blue-white. Film manufacturers set standards that established daylight at 5500°K, and daylight color films were balanced to that color temperature. Tungsten illumination, however, is set at 3200°K for type B color films (35mm color negative) and at 3400°K for type A color films (Kodachrome and Ektachrome). For filming, daylight color temperature is a combination of direct sunlight and reflected skylight. This combined light varies greatly with the time of day and the atmospheric conditions; for example, overcast, hazy, partly cloudy, or full sun.

Color Temperature Meters

If you want to know exactly the color temperature of a source light, you have to use a color temperature meter. The simpler meters register the relationship of the red and blue portion of the illumination; they give the temperature in degrees Kelvin or convert it to show the correction filter required for the particular color film being used. More sophisticated color temperature meters, which include mini-computers in their circuitry, not only show the color temperature in a direct reading but also indicate the filter or filter combination required for the correct light balancing. These types are known as *two-color meters* because they read only the red and blue portion of the light. They cannot be used to read fluorescent illumination, which does not emanate from a heated black body, or filament.

The three-color temperature meter made by Spectra reads fluorescent light because it reads light on two different scales; the red/blue content and the red/green content of the light source. With conversion charts, it indicates the combination of filters required for correct light balancing. Minolta's color temperature meter also reads fluorescent light, and it offers a digital read-out of the correction filters required for balancing various color emulsions. Three-color temperature meters can also be used to read directly the Kelvin rating of daylight, tungsten, arc, or metal halide (HMI) light sources. They offer the greatest versatility.

CONVERSION, BALANCING, AND COMPENSATING FILTERS

There's often confusion about the names of the different filters that are specifically designed to affect the color sensitivity of the color emulsions.

Conversion filters are the basic filters used to balance tungsten film to daylight. The amber filters, #85 and #85B, shift the color temperature from 5500°K to between 3200°K and 3400°K. The blue #80 series, used for daylight color film under tungsten lights, shifts 3200°K to between 4200°K and 5500°K. The color shifts are dependent on the exact Kelvin rating of the light source.

Light-balancing filters shift the color temperature of the light source in small degrees to balance the color of the light to the color sensitivity of the film. This series of filters is primarily used with reversal color films or when the cinematographer wants to compensate for a lens with a color response different from others being used. These filters are the #82 series (bluish) and the #81 series (yellowish).

Color-compensating filters act as "light sponges" in that they absorb excess amounts of undesirable color from light sources. A color-compensating (CC) series is available in six increasingly absorbent densities of yellow, magenta, cyan, red, green, and blue. Each filter attenuates its complementary color a specific degree.

The filter charts in Tables 1 to 3 more clearly demonstrate how these filters work.

Conversion Filters

Because of the differences between light sources, film manufacturers created one color emulsion balanced specifically for daylight illumination and an-

Table 1 Kodak Color-Conversion Filters for Color Film

Filter Color	*Filter Number*	*Conversion in °K*	*Exposure Increase, in Stops*
Blue	80A	3200–5500	2
	80B	3400–5500	1-2/3
	80C	3800–5500	1
	80D	4200–5500	1/3
Amber	85	5500–3400	2/3
	85B	5500–3200	2/3
	85C	5500–3800	1/3

Author's note: Originally, most color film was balanced to photoflood illumination (3400°K), and the #85 filter was used for filming in daylight. Old habits are hard to break! Although the #85B is the correct conversion filter for tungsten (3200°K), the old #85 is still in wide use, and the film labs make the slight additional correction as a matter of course.

Table 2 Kodak Light-Balancing Filters

Filter Color	*Filter Number*	*To Obtain 3200°K from*	*To Obtain 3400°K from*	*Exposure Increase, in Stops*
	82C + 82C	2490°K	2610°K	1-1/3
	82C + 82B	2570°K	2700°K	1-1/3
	82C + 82A	2650°K	2780°K	1
Bluish	82C + 82	2720°K	2870°K	1
	82C	2800°K	2950°K	2/3
	82B	2900°K	3060°K	2/3
	82A	3000°K	3180°K	1/3
	82	3100°K	3290°K	1/3
No Filter Necessary		3200°K	3400°K	
	81	3300°K	3510°K	1/3
	81A	3400°K	3630°K	1/3
Yellowish	81B	3500°K	3740°K	1/3
	81C	3600°K	3850°K	1/3
	81D	3700°K	3970°K	2/3
	81EF	3850°K	4140°K	2/3

other balanced specifically for tungsten. Filter manufacturers then created color-conversion filters, which enable cinematographers to use a daylight-balanced film under tungsten light. A blue filter, #80A, changes the mostly yellow color of tungsten light to a bluish color and brings it into balance with daylight film. However, because the exposure must be increased by two stops to compensate for the light lost through the filter, it's not very practical except in an emergency. A tungsten-balanced film can be used in daylight by adding an amber conversion filter, #85 or #85B. Although there's a light loss of approximately two-thirds of a stop, there's usually enough light outdoors to make the exposure compensation insignificant. Even in low light, it's usually manageable.

The general practice is to use a tungsten-balanced color stock for both tungsten and daylight filming. The 35mm color emulsion most frequently used, Eastman #5247, is balanced to tungsten illumination (3200°K). This emulsion is used for

- interiors with tungsten light
- night exteriors with tungsten light
- day interiors with "blue" (daylight + HMI) light with a #85 or #85B conversion filter
- day exteriors with a #85 or #85B conversion filter

New emulsions balanced for daylight that require no conversion filter have recently become available, and cinematographers are beginning to use them for exterior filming (Eastman #5245 and #5297, Fuji #8520). New "high-speed" emulsions (Eastman #5296, Fuji #8514) are being used more and more for interior tungsten filming and night exteriors; however, due to their high degree of light sensitivity, they are not generally used outdoors in daylight with a conversion filter because either a heavy neutral density filter must be used or a very small lens stop is necessary to control the exposure, making it very difficult to see the ground-glass image. Whatever emulsions you decide to work with,

Table 3 Kodak Color-Compensating Filters

Peak Density	*Yellow (Absorbs Blue)*	*Exposure Increase, in Stops*	*Magenta (Absorbs Green)*	*Exposure Increase, in Stops*	*Cyan (Absorbs Red)*	*Exposure Increase, in Stops*	*Red (Absorbs Blue and Green)*	*Exposure Increase, in Stops*	*Green (Absorbs Blue and Red)*	*Exposure Increase, in Stops*	*Blue (Absorbs Red and Green)*	*Exposure Increase, in Stops*
0.05	CC05Y	——	CC05M	1/3	CC05C	1/3	CC05R	1/3	CC05G	1/3	CC05B	1/3
0.10	CC10Y	1/3	CC10M	1/3	CC10C	1/3	CC10R	1/3	CC10G	1/3	CC10B	1/3
0.20	CC20Y	1/3	CC20M	1/3	CC20C	1/3	CC20R	1/3	CC20G	1/3	CC20B	2/3
0.30	CC30Y	1/3	CC30M	2/3	CC30C	2/3	CC30R	2/3	CC30G	2/3	CC30B	2/3
0.40	CC40Y	1/3	CC40M	2/3	CC40C	2/3	CC40R	2/3	CC40G	2/3	CC40B	1
0.50	CC50Y	2/3	CC50M	2/3	CC50C	1	CC50R	1	CC50G	1	CC50B	1-1/3

These filters can be used singly or in combination to complete almost any desired correction. The CC filters are used primarily for color transparencies (stills) or for 16mm reversal material that isn't going to be duplicated, thereby eliminating the opportunity for post-production correction. Light sources can vary greatly and depend on the age of the lamp, the voltage applied, and the lamp reflector and housing. Outdoors, daylight can also vary greatly due to atmospheric conditions—haze, smog, clouds—and the time of day. In 35mm professional cinematography, the color compensations offered by using these filters are easily accomplished by the lab during the answer print process, obviating the need to use them during the original filming.

it's a simple matter to learn the correct color-conversion filter to use.

If a #85 or #85B filter is accidentally smashed, lost, or forgotten, you can adjust the exposure to compensate for the lack of the filter and ask the lab to color-correct the print. The results will be excellent. Similarly, if you must film with available fluorescent light and you don't have the correct conversion filter, or if there is such a low level of light that you can't use the #85 and get sufficient exposure, shoot the scene without a filter and let the lab correct the print. Why even bother with the #85 filter? With it, the film lab enjoys a wider range of possible color corrections. Without the #85, the lab uses most of its ability to correct the color to get a normal print, and if any further corrections are called for, the lab might not be able to accommodate the request. If you must film without the #85 or #85B, be sure to notify the lab, and indicate on the camera report those scenes that require special handling! They need not be on a separate film roll. (See Color Plate 10.)

Light-Balancing Filters

Light-balancing filters comprise two series of two different colors. One set is bluish and is available in increasing degrees of intensity. The blue filters balance tungsten color negative to tungsten light, which may be slightly redder or may have a lower color temperature due to the time of day, voltage variations, old bulbs in bad reflectors, or other conditions. The blue color filters out the excessive red of the illumination to bring it effectively back to a 3200°K rating. Outdoors, as the combined daylight gets redder toward sunset or in the very early morning, the blue light-balancing filters can be used in addition to a color-conversion filter (#85 or #85B) to keep the light source in balance with the film. When working with 35mm color negative, the general practice is to use only the color-conversion filter, leaving the more subtle adjustments for the film lab to make through their printing procedures. However, if you are working with 16mm color reversal film, such as Kodachrome or Ektachrome, you may need as much color correction as possible while you are shooting; in this case, you would add a light-balancing filter.

The other set of light-balancing filters is amber and also has increments of intensity; the choice is made on the basis of the degree of correction desired. These amber filters are used to reduce the color temperature of the light source and to bring it into balance with the color emulsion. For example, when filming under overcast conditions and the prevailing light has a high Kelvin rating, you can use one of the amber filters to great advantage. Light-balancing filter charts show exactly which filter to use for various degrees of light balancing. Remember though that when using color negative emulsions, as opposed to color reversal film, slight variations in the light source can easily be corrected later by the film lab. On heavily overcast days I often use a coral #1 or #2 in addition to the #85 to warm the overcast light. This incorporates the correction into the negative and gives the dailies a better look. Coral filters are discussed later in this chapter.

Both series of light-balancing filters have subtle hues that bring color film into balance with various light sources that might be slightly different.

Color-Compensating Filters

Another category of color filter is used primarily for its color-shifting ability: color-compensating filters. These filters come in different densities of red, green, blue, cyan, magenta, and yellow. They compensate for any deviation of true color reproduction by any of the elements involved in creating a true color image. These variables are often overlooked by cinematographers because labs can easily compensate for any balance problems.

The one instance in which some directors of photography prefer to use color-compensating filters is when using certain lenses that have a slightly different color response. When testing lenses, cinematographers may discover that the image produced by a particular lens or perhaps a whole set of lenses of different focal lengths is slightly colder (bluish) or warmer (yellowish) than normal. To obtain dailies in color balance, a color-compensating gelatin filter can be adhered to the back element of that particular lens. The slight color shift of the lens could also be corrected later in the lab when color correcting or timing the print. Camerapersons using 16mm reversal film, as well as still photographers, use color-compensating filters more than theatrical cinematographers do because their original color film and transparencies are used directly for reproduction and lab control may be limited.

Fluorescent Light Filters

Fluorescent lighting presents a special problem for cinematographers. Directors of photography often find that it's far simpler and more expedient to use available fluorescent lighting than to try to light the set with other kinds of lights. When film locations are very large, it's much easier to use a little supplementary lighting on foreground subjects than to relight the entire area. Imagine the complex job of lighting airport terminals, train or bus stations, large office complexes, hospitals, or supermarkets!

The fluorescent tube contains a chemical coating on its inner surface that interacts with the voltage entering at both ends of the tube, causing the chemical to glow (fluoresce) and thereby create light. Because it is not a black body heated to emit light, the fluorescent tube has no Kelvin rating. Its spectral distribution differs greatly from tungsten light. Color film response to fluorescent light is very different from that of the human eye: there's a high level of green light in a fluorescent tube that humans cannot discern. By referring to charts made by filter manufacturers and using the recommended filters, all color films can be brought into balance with fluorescent light. Some color temperature meters read the output of fluorescent tubes and indicate the filters needed for light balancing; for example, the Minolta color temperature meter and the Photo Research Spectra three-color meter.

The color-conversion filter that has been designed to balance the source of cool-white fluorescent light to tungsten type B (3200°K) film is the FL-B filter (*FL* stands for fluorescent, and *B* indicates type B color film). Another filter, the FL-D, balances cool-white fluorescent light to daylight film. Any additional lights used in conjunction with fluorescent lights must be color-balanced by putting filters on those supplementary lights. Rosco and Lee both manufacture gel sheets that can be used to bring various light sources into balance with fluorescent tubes. A fluorescent tube called *Optimum 3200°K* equals the color quality of tungsten light and does not require a filter. When using daylight film (or 3200°K film with a #85 filter) and there's a combination of daylight and fluorescent light, the Chroma-50 fluorescent tube is correctly balanced to match the daylight. HMI lights can be used with Chroma-50s for any additional supplementary lighting. Another way to control fluorescent light is to gel the fluorescent units themselves.

With these many options for balancing lights and film, the cinematographer must decide which method best suits the situation at hand. If there are only a few overhead fluorescent units, it's generally simplest to gel those units and to use supplementary tungsten or HMI lights. If very low light levels make it impractical to filter the camera and all the light within the scene is the same (or color-balanced by using the same kind of fluorescent tubes for fill light as the overheads), then the film lab can often do all the color correcting that's necessary to produce a normal color print. (See Appendix: Rosco.)

To review: the color-conversion filters, light balancing filters and color-compensating filters, all have their individual capabilities. By studying the filter charts, you can get a better grasp of their names, numbers, and various uses.

The color response of the color negative can be shifted in any direction through the use of a color filter to modify the light striking the color emulsion. The discussion of color filters has been limited thus far to shifting the color response of a particular color film to match that of the available light source and bringing an out-of-balance situation into correct balance to produce a normal-looking print. The color response of the color negative can also be shifted to create a special look.

CORAL FILTERS AND SEPIA TONE

One of the most popular color filters is the coral filter, which has a warm brownish amber tone. Its primary function is to add the warm atmosphere of candlelight or firelight to a scene. The degree of warmth that is added is controlled by the intensity of the filter used. Coral filters are available in increments from #1 to #12, and filter manufacturers make heavier grades and graduated coral filters on request.

Similar effects on lighted interiors or exteriors can be achieved by adding warm-toned gels, such as color temperature orange (CTO) ¼, ½, or full, to the lights. Perhaps surprisingly, not all the lights need to be gelled. The key light can remain "white" while the back light and effect lights are gelled to represent the candlelight or firelight. A coral filter can also be used to add an overall warm atmosphere. There are more varieties of coral filters than there are gels of the same tonal value, and a combination of the two usually works very well.

To heighten the effect of a real sunset, cinematographers often add more warmth to the scene. A coral #5 (C-5) is equivalent to a #85 color-balancing filter. Two #85s are often used together for sunsets, and a C-10 can be used by itself for the same effect. Some cinematographers carry a specially made "double 85" filter for sunset effects. On the other hand, a regular #85 with any coral filter gives the director of photography complete control over the degree of warmth added to the scene. A graduated coral-to-clear or #85-to-clear filter is often used to redden the sunset sky without affecting the foreground color. To continue filming after the sun has set, coral filters, and a lamp or two can help get that one last matching shot to finish the sequence.

Coral filters are commonly used to film on overcast days when normal daylight is predominantly blue or to match a scene being filmed in overcast to one that was shot earlier in sunshine. This correction can be made to some degree in the lab, but the cinematographer cannot always be around to time the final print. Suppose the prevailing overcast light is about 12,000°K, according to the color temperature meter. By putting a C-3 filter on the lens, you can correct the light source to about 5900°K, and the #85 filter corrects it close to 3200°K, which puts you in balance with type B color film. (See Color Plate 11.)

A word of caution: If you're using supplementary lights, make sure that they match the prevailing light source or the coral filter will accentuate the color difference. Simply stated, an HMI or arc light will match sunny daylight of 5600°K. If, however, the prevailing daylight is slightly overcast and has a color temperature of 9000°K or higher, you should filter your supplementary lights by placing a ¼ or ½ color temperature blue (CTB) gel filter on the lamps, thereby increasing their color temperature to match that of the overcast, and then correct both together with a coral filter on the lens. Although a color temperature meter can be used to match supplementary light to available light, you'd be amazed to learn just how closely you can judge this match by eye if you train yourself to be extremely objective when looking at the color of supplementary light compared to available light. The only circumstance in which matching by eye isn't feasible is when working with fluorescent light sources because there's no correlation between what the eye sees and what the color film "sees."

While on the subject of balancing supplementary light to available light, a similar situation exists during early sunny daylight hours and late afternoon hours. At these times, CTO gels can be used to lower the color temperature of HMI, arc or daylight blue FAY lamps, and to balance those lamps to the warm orange daylight source. Rosco in New York and Hollywood publishes a very comprehensive booklet about lamp filters.

It's always a good idea to note on your lab report exactly which filters you used on the camera. This has a double advantage. First, the lab knows that you have a particular effect in mind, and second, knowing that a filter was used, the lab will not mistakenly try to eliminate the effect thinking it was accidental. Unless you instruct the lab otherwise, they will always try to correct the print to produce a "normal" look. An added advantage of recording the filter is the ability to refer to the record after viewing the dailies or in the event that an additional matching scene must be shot or a retake is required.

Coral filters can also replicate the look of old-time photographs, which were made in sepia, or brownish tones. Sepia tone is often used for flashback scenes. The tone is controlled through the selection of the coral filter and the degree to which exposure is or is not compensated. Denser coral filters remove more of the blue and green portions of white light and thereby create the sepia tone. Coral filters have a filter factor needed to compensate for light lost when they are used, therefore, the exposure must be adjusted to compensate for that loss. After you have filtered and exposed for sepia tone, ask the lab to help with additional color correction if you want a deeper sepia tone. (See Color Plate 12.)

Some Guidelines for Making Sepia Tone Scenes

Making a sepia tone effect in daylight with a tungsten-balanced film requires the normal #85 filter *plus* those listed below. When using a daylight-balanced film in daylight, no #85 is used, but the filters listed below create the effect.

- An overcast day is best when colors are more muted.
- Try to avoid brilliant colors, particularly blues and greens.
- Use a heavy coral filter, #4 or #5 (#5 equals an additional #85).
- An additional "Chocolate" (Rosco Cinegel #682) filter to reduce blues and greens and add more brown tone (see Chapter 5).
- Add a regular fog #3 to soften the sharpness and add an "old-time" feeling.
- If the scene is of normal brightness, underexpose one-half stop.

A film I shot recently, *The Neon Empire*, covered the period from the mid-thirties to the mid-forties. The story was set in Las Vegas, and the director wanted a hot desert feeling. I used filters to help achieve this goal. The exteriors were filmed with a coral #3, a regular fog #3, and a normal #85 filter, and were underexposed by one-half stop. The interiors were filmed with a coral #2 and a regular fog #2, again underexposing by one-half stop. The underexposure eliminated the slight flare effect of the fog filters and enhanced the coral effect, and the combination of stronger filters made the exteriors appear hotter. Experienced camerapersons know that tests are necessary to define the filter and exposure combination for any desired effect.

NEUTRAL DENSITY FILTERS

Aside from color filters, there is another widely used series of filters: neutral density filters. The ND filter has an overall gray tone—no color at all—therefore, it neither affects the color balance of the color negative nor in any way changes the color response. The ND filter is strictly for control of exposure when the lens diaphragm alone is not enough, or if you want to film at a wider lens opening to control the depth of field. For example, suppose the brightness of a ski slope requires an aperture of F/32 and your lens only stops down to F/16, or if the normal exposure is F/11 and you want to film at F/5.6 to reduce the depth of field. A neutral density filter solves the problem.

Neutral density filters are marked as .30, .60, .90, and 1.20ND. You can use formulas to calculate the light absorption of each filter and thereby the decrease in exposure; but it's simpler to remember the following:

.30ND reduces exposure by one stop.
.60ND reduces exposure by two stops.
.90ND reduces exposure by three stops.
1.20ND reduces exposure by four stops.

Neutral density filters can be used in very imaginative ways. Controlling the contrast of an image by keeping the lens diaphragm open was discussed earlier. Using a neutral density filter is the way to do just that. If the normal exposure is F/16, the same scene filmed at F/5.6 will be less contrasty. The difference is three stops, so a .90ND will do the job perfectly. Suppose the background of a close-up or medium shot appears so sharp due to the depth of field that it is intrusive. By keeping the lens more open, the background will be softer and more out of focus, and the separation between the sharp foreground and the soft background will create a greater three-dimensional effect. The 1.20ND allows you to open up four stops and shoot at F/4 instead of F/16.

The neutral density filter can be a timesaver too. If you have many scenes to shoot indoors with high-speed emulsion (Eastman #5296) and only one or two shots to be made outdoors in sunlight, a .60ND will reduce the amount of light that reaches the fast film stock by two stops. This makes it virtually equal to the film speed of Eastman #5247 (with a #85 filter), which would normally be used for daylight exteriors. There are three benefits: a film magazine does not have to be prepared, the camera does not have to be reloaded, and the production team does not have to carry two different raw stocks. If you have many exterior daylight scenes, it would of course be better to reload with the slower film stock.

Because the neutral density and #85 and #85B filters are so widely used outdoors in daylight, filter manufacturers have combined the two types of filters into one. The #85N3, #85BN3, #85N6, among others, are readily available. They offer the advantage of not having to place two glass filters in front of the lens or use a double gel filter behind the lens, which is not advisable. H&H and Tiffen make combination glass filters, and Eastman manufactures the gel filter combinations.

GRADUATED FILTERS

A type of neutral density filter that offers additional image control is the graduated neutral density filter. About half of the filter area is clear glass, and the other half is an ND or an ND combined with a color. Other graduated filters are various color-to-clear filters, such as #85-to-clear. The line where the ND or color and the clear glass interface can be a very soft edge to a hard edge. Both types of edges have particular advantages. Cinematographers sometimes want a sharp delineation, and other times a very soft, unnoticeable blend is better. It often depends on the foreground subject and just how dramatic an effect is desired. (See Color Plate 13.)

The ND-to-clear filter is primarily used to hold back the usual overexposure of the sky if it isn't a true blue. (These days, when smog abounds, we don't often have a true blue sky unless we've traveled away from urban areas.) By looking through the camera, the cinematographer can place the edge of the ND along the horizon so that the sky receives less exposure than the foreground, which is filmed through the clear part of the filter. No increase in exposure is required in this instance. The normal exposure for the foreground filmed through the clear half of the filter will be correct, and the sky will be darkened proportional to the filter factor of the ND. (See Color Plate 14.)

A word of caution: The final positioning of the filter must be viewed on the ground glass with the lens stopped down to the correct F/stop to determine if the line of demarcation between the ND and the clear glass will be visible. If visible, the line may be objectionable. This is particularly true when a wide-angle lens that has great depth of field is used. If the line does show, open the diaphragm until the line becomes so far out of focus that it's almost invisible, and use an overall ND filter to correct the exposure. The closer the graduated filter is to the lens, the easier it is to hide the blend; however, it is also more difficult to see just where the blend line is. If you want a virtually invisible division line, make certain that you request a "soft blend" when renting filters or ordering them for purchase. (See Color Plate 15.)

Detection of the blend line is affected not only by the distance from the filter to the lens itself but also by the focal length of the lens. Lenses with a shorter focal length—18mm to 40mm (on 35mm cameras)—make the blend line easier to see. Because graduated filters are generally used in wide-angle scenes with large areas of sky, snow, water, or other areas to be toned down or colored, be especially careful when setting the lens stop.

It's very difficult to use a graduated filter with lenses of a long focal length. To see the exact location of the blend line, the filter must be relatively far from the front of the lens. Fortunately, graduated filters are rarely used with long lenses because there is little sky or other area to be filtered. As mentioned previously, telephoto lenses cannot accommodate any glass in front of the lens.

When using a graduated filter on a zoom lens, you must exercise additional caution. The filter will probably be used for a wide-angle setting. If you don't zoom or move the camera, there will be no problem; however, if you zoom in, you must observe two things. First, how does the filter look when zoomed in? Second, if the camera pans or tilts when zooming in, will the graduated filter become apparent due to the camera movement?

Graduated filters are often oversized so they can be shifted for correct positioning in an adjustable filter holder. Check with the camera rental house if you need an adjustable filter holder. With the exception of wide-angle lenses, there may be enough shifting room within the normal filter holder and matte box. The newest matte boxes are designed to work with oversized filters. One or more sections of the matte box rotate so that a graduated filter can be adjusted to conform to an off-level landscape line. (See Color Plate 16.)

When the sky is gray or void of color, you can add color several ways. One way is to use a graduated #85-to-clear with a 60ND-to-clear. Place the #85 over the lower foreground portion of the scene. Because the upper section is clear and the tungsten-balanced film is being used in daylight, the light from the sky will record bluer on the film, but it will also be overexposed. The graduated .60ND filter placed over the sky area will reduce the sky brightness by two stops. This effect can be achieved with a combination single filter, such as #85-to-.60ND. Another approach to the same situation is to use a graduated blue-to-clear filter (B5-to-clear) over the sky area in combination with an overall #85. The graduated blue filter will tint the sky and reduce its exposure by adding more blue. Your exposure should only compensate for the overall #85.

Graduated filters give cinematographers an unlimited variety of effects. H&H and Tiffen make up filters in any combinations or colors requested. Graduated filters are by no means limited to creating sky effects or to controlling sky exposure. Many scenes shot on a white beach or ski slope would benefit from a graduated ND filter arranged with the clear part at the top, or at an angle and the ND half over the bright sand or snow. Also consider a scene where half of the street is in bright sunshine and the other half is in shade. You can use the graduated ND sideways with the clear area over the shady side of the street. Graduated filters can be rented or purchased in the new 6.6" × 6.6" size with an accompanying matte box.

When working with any graduated filter—ND or color—be extremely careful about camera moves, and watch for actors who walk through the blend line! Fortunately, most scenes that require a graduated filter are long shots, and the camera may not move, allowing the action to be contained below or above the blend line. Graduated filters used for sky control usually allow some leeway for panning the camera. Tilting up or down, however, can be a problem.

Some filters have a rather sharp line of demarcation between the clear area and the ND or color section, and others are blended more softly. Filters with a soft blend can be used more easily, and the line is not as apparent when the lens is stopped down. Cinematographers can sometimes make very good use of a sharp separating line. When using a lens stop sufficiently large to keep the sharp line somewhat out of focus, these graduated filters can be used to create a dramatic glow along the horizon line that is similar to the dramatic sunsets or stormy skies created by nature. (See Color Plate 17.)

NEUTRAL BLENDED RATIO ATTENUATORS

The neutral blended ratio attenuator (NBRA) is similar to the graduated filter but is more sophisticated in the way it works. The NBRA, commonly called an *attenuator*, is currently made in neutral densities and is, in a sense, a graduated filter except that there's no blend line. The one-stop NBRA has a .60ND at one edge of the filter, and the opposite edge is clear. Starting at the clear edge, the neutral density increases so gradually that there is no line of demarcation. The blend is completely continuous. At the midpoint, the neutral density is .30ND; therefore, exposure compensation of only one stop is needed.

This is how the NBRA filter works. Imagine a normal scene with green shrubbery in the foreground, a house in the middle area, and sky and clouds above. The one-stop NBRA, positioned in the matte box with its densest area uppermost, offers the following control: Compensating for the one-stop filter factor gives the house a normal exposure. The sky falls in the upper .60ND area, thus its exposure is reduced by one stop (.60ND is two stops and the exposure was already compensated by one stop for the filter), making the sky darker by one stop. The foreground, which is dark green, gets the additional exposure of almost one stop because the bottom edge of the attenuator is clear! The movement of people within the frame is usually no problem, as long as they don't travel from one extreme of the attenuator to the other. Similarly, most camera moves don't create a problem when an attenuator is used. The attenuator, like the graduated filter, can be used with the densest section on the top, bottom, or side, or at an angle if you have a rotating adjustable filter holder in your matte box.

Because there is no blend line in the NBRA, even a wide-angle lens can be used at a small lens stop to obtain extreme depth of field. The NBRA is a versatile tool for controlling the color image. Attenuators are available in ratios of 1, 1½, 2, 2½, and 3 stops. They are made by Pancro-Mirror, which specializes in making neutral density filters by its own patented technique. For what it's worth, I use their filters exclusively for all my neutral density requirements.

POLARIZING FILTERS

The polarizing filter looks very much like a neutral density filter; however, its use is completely different, and it's always round. The polarizing filter is neutral in color and requires increased exposure; its filter factor is usually two stops, depending on the manufacturer. Although its application is quite varied, it's primarily used for two controls: to darken a weak blue sky, which will accentuate clouds, and to diminish unwanted reflections.

The trick is knowing when the polarizer will work and when it won't! Polarized light is light that is reflected at certain angles from the sky and off of any shiny surface. Imagine a single light beam from any source. This ray of light vibrates in a multidirectional path around its axis until it is reflected at certain angles from a polished surface or reflected back from the sky at an angle that's about 90° from the sun's position. The reflected light beam vibrates in only one plane in relation to its axis. The polarizing filter is, in a sense, made up of invisible parallel lines, much like a Venetian blind. These lines don't open or close, but they can be rotated, and all polarizers are round for this reason. If the beam of light is reflected at a particular angle from a polished surface, it becomes polarized so that the light vibrations are in a plane that's horizontal to the axis. If the polarizing filter on the lens is set so that its invisible screening lines are also horizontal, all of the polarized light from the beam will pass through the filter. If the filter is rotated so that its screening lines are perpendicular to the reflected polarized beam, it will not permit those particular rays of light to pass. (See Figure 4.)

The effect of the polarizer is easily seen while looking through the filter and rotating it. When the unwanted reflections are diminished to the proper extent, or eliminated or when the sky is darkened sufficiently, the filter is marked and put in the matte box in the same relative position. The polarizer affects only polarized light. All other light beams pass freely through the filter, and because it is a neutral

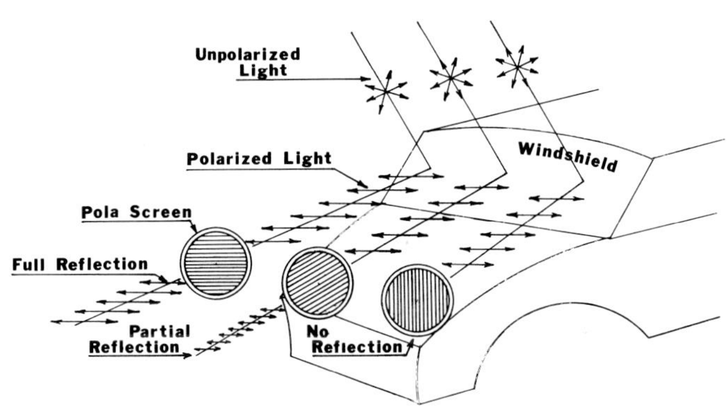

Figure 4 How the polarizer (pola screen) works.

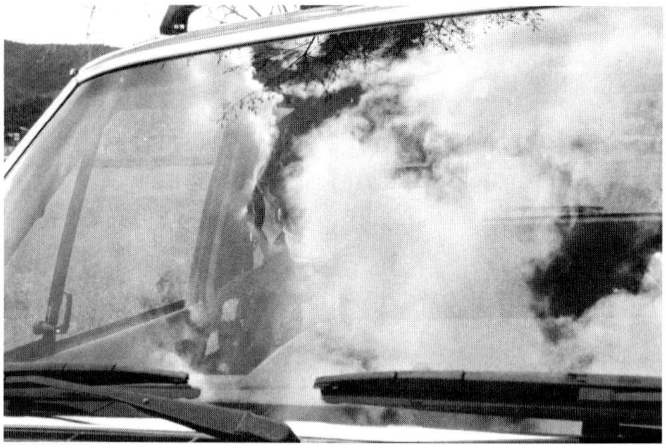

5A

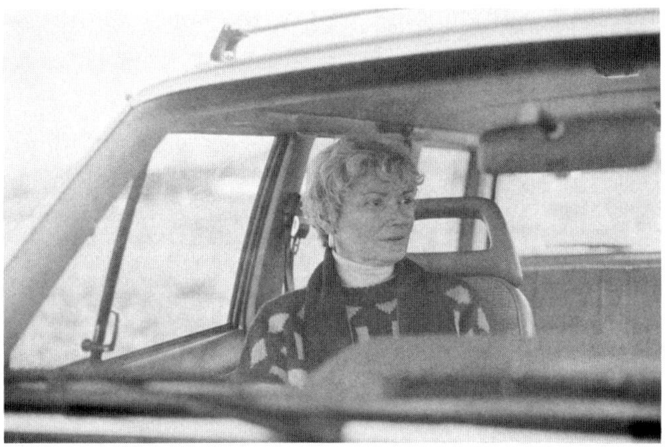

5B

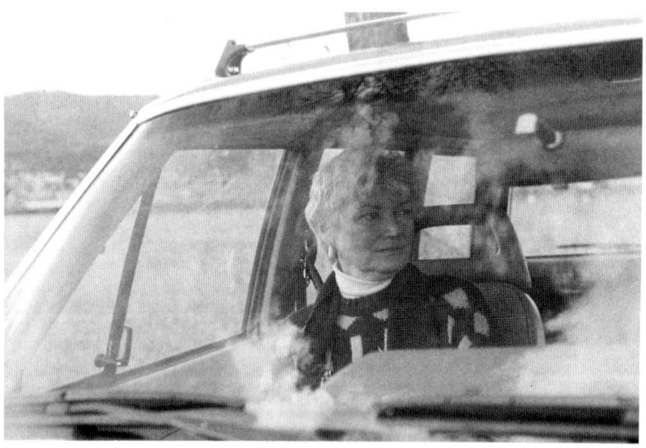

5C

Figure 5 Shooting through a windshield is often a problem due to reflections, as shown in Figure **5A**. In Figure **5B**, the polarizer fully eliminated the reflections, but this almost makes it appear as if there is no windshield. For Figure **5C**, the polarizer was partially rotated until some reflection could be seen but the actor was also fairly clear.

color, it has no effect on colors other than to make them more saturated if their hue is diminished by the polarized light. In motion pictures, the polarizing filter is often used when filming from the hood of a car through the windshield. Reflections of the sky and clouds on the windshield can all but block out the people inside. With a polarizer, these reflections can be eliminated or reduced sufficiently to allow the camera to see through the glass. Be careful when using the polarizer in this situation. If used to its fullest capacity, the polarizer can make the windshield completely invisible, and the shot will look unrealistic. Reduce the effect of the polarizer by rotating it, and some reflections will reappear. As the car travels, the images of overhead trees, clouds, and sky can partially superimpose themselves over the people inside the car, providing a more realistic shot. (See Figure 5.)

The polarizing filter is also used to darken an otherwise pale blue sky to make clouds stand out better. You are limited in this use because the light from the sky becomes increasingly polarized only as it approaches an angle that's 90° from the sun's position. This area reflects the maximum polarized light and is therefore the most filterable. The polarizer's effect in darkening the sky is slightest nearest the horizon, and the effect increases as the camera tilts upward. As the sky becomes less blue nearer the sun, the darkening effect also lessens. The effectiveness of the polarizer in darkening the sky is easily judged by looking through the filter and rotating it to find the position that offers the desired polarizing effect. Caution: When using this filter, if the camera pans far enough, the view can change from polarized skylight to unpolarized skylight, and the color of the sky will change during the shot.

A sky that appears whitish or gray cannot be darkened by a polarizer. The polarized light that is normally reflected from a blue sky is being blocked by an atmospheric condition of moisture, haze, smoke, dust, or slight overcast. In this instance, the cinematographer must find another way to control the sky, for example, a blue-to-clear graduated filter, an ND-to-clear filter, or both. However, these substitute measures are not as effective. Using the polarizer, people are not affected by the filter and are free to move about within the frame, which is not the case with graduated filters.

Because the polarizer is neutral in color, it does not affect any part of the color spectrum; however, due to its density, it does require an increase in exposure of about two stops, depending on the manufacturer. You can easily identify the correct exposure by taking a reading with a spot meter through the filter.

Polarized light is reflected from most shiny surfaces. Glares that appear from furniture, windows, water, sky, and so on, can usually be reduced sufficiently to make them much less objectionable. Filming through the surface of water to objects below is also helped tremendously by a polarizing filter.

By reducing glare, the polarizer increases color saturation, and this may be one way to increase contrast when needed. As always, view the effect on the ground glass to see if you are screening out too much reflection. Also check any instance where the camera pans from a polarized area to a nonpolarized area. If you use a reflectance-type exposure meter, which reads the subject plus the reflection or glare, and you then eliminate those reflections, you must adjust your exposure reading. The most accurate exposure reading when using a polarizer for windshield shots, is to take the reading through the filter, in the correct rotated position, with a spot meter pointed at the actor. You can also read the incident light and compensate for the polarizer's filter factor.

Cinematographers try to create images that heighten and enhance the dramatics of the story. These images are realized most successfully when they emerge from the work accomplished on the set. The effects created on the original negative are usually far superior to those created in post production unless they are special optical effects. The lab's greatest contributions are made when processing the original negative according to the special requests of the cinematographer and later when fine-tuning all the effects during the timing of the final release print.

Some of the controls discussed thus far that affect the color negative could be duplicated to some degree by the film lab. These controls are mainly in the realm of color correcting the print to eradicate differences in available light sources or to make a print appear warmer (more amber) or cooler (more bluish) in tone. However, the effects of using a polarizer, graduated filter, fog filter, or low-contrast filter could never be duplicated in the lab or in post production. These are the cinematographer's on-the-job tools for controlling the image.

TESTING AND LEARNING MORE ABOUT FILTERS

Learning how to use filters is not an easy job, but like so many things, practice and experience are the best teachers. The first step is to test your filters under various conditions. These preliminary tests can be done with either stills or motion pictures. Because stills are less complicated and less costly, they're a good choice.

If you're ultimately going to use 35mm motion picture color negative, you'll want to use that emulsion for your tests because different emulsions can and will react differently to different filters. In California, Oregon, and other states, some color labs sell and process 35mm cassettes of popular professional emulsions. RGB Color Lab handles Eastman #5247, #5296, #5245, and #5297. Evergreen Film Service processes Agfa #XT-125 and #XT-320.

RGB Color Lab	Evergreen Film Service
816 No. Highland Ave.	1412 West 7th Ave.
Hollywood, CA 90038	Eugene, OR 97402
Tel.: (213) 469-1052	Tel.: (503) 686-2183

The processed color negative can be printed on paper or made into color slides. For testing purposes, it's wise to request that all of your prints or slides be printed on one light in the middle of the lab's printing scale, otherwise you will have no way of knowing what the filter does in regard to over- or underexposure. RGB and Evergreen process all Eastman, Fuji, and Agfa emulsions normally but also offer additional processing at two or four times the ASA rating (pushed one or two stops). You can request mailer envelopes if you live out of town. The service by mail is very good.

The first series of tests will familiarize you with

how the different filters react under different light conditions. As an example, select an exterior location on a sunny day for a contrasty shot—perhaps a white house or church that has deep shadow areas or is surrounded by dark foliage. Bring along a gray scale, a color chart, and a slate on which to note the filters used and whether the exposure is normal, under, or over, and by how much. Set your camera on a tripod, and before using any filters, take a careful exposure reading and manually set your camera accordingly, by-passing any automatic exposure controls. Shoot the first frame clean, that is, with no filter, for comparison. Then proceed to the first filter, shooting at one-half and one full stop under, then normal exposure, then one-half and one full stop over. In this way, you'll learn not only what the filter does but also how exposure affects it. The gray scale, color chart, and slate should be included in each frame.

Follow this method for all of the filters of the particular series you're testing. A contrasty situation is fine for testing regular fogs, double fogs, and low-contrast filters. An overcast day allows you to discover the effects of the coral filter series and the regular and double fogs. You may add diffusion filters (see Chapter 5) during any of these tests to see the combined effect. If you have friends who'll pose for you, see what happens when you film a person against a bright sky or when wearing a white shirt and standing in front of very dark foliage. You can test the filters in any situation to see what they will do, and the more tests, the better! Try situations in which the sun comes from over your shoulder in a relatively flat light; then move around so that half of your subject's face is brightly lighted and half is in shadow; finally, use the sun as a back light. Look at the gray scale and color chart critically to see how they too are affected by the various filters.

Through testing and examining the results, you'll soon learn a lot about filter effects, but that's only the beginning. Try to make the filters do the work of creating the best possible picture without lighting equipment or reflectors. You'll then be ready for the next step.

Choose a filter that you feel is best for a particular subject or scene, then see how using the same filter for a long shot, a medium shot, and a close-up affects the image. Is the contrast equal for all three? Would a filter of lesser or greater strength be better for the medium shot or the close-up? Each setup can require a filter of different intensity to create a matching look when the scenes are placed side by side. For example, imagine a scene in which two people stand facing each other, with a white church surrounded by trees in the background. It's a bright day, and the sunlight strikes the scene with a three-quarter front light from the left. The first shot is a loose full figure of the people with the white church behind them. Choose the filter you think will help reduce the contrast without resorting to lighting equipment. To cover the scene, you first want a close-up of the person on the right. When you set up, you notice that the background is no longer the white church, but dark green trees, and the sunlight is striking the face almost head-on. Do you use the same filter? Possibly not, but try it and try other filters which may make the close-up a better match to the full figure shot when you see the final results. Shooting without any filter might create the best match. When the camera is turned around to photograph the person on the left, you see that the subject is in back light and has mostly sky as a background, a situation very different from the other two setups. Again, try different filters to see the results. Just by thinking about the above example, you've already begun to appreciate the complexity of using filters accurately.

UNDERSTANDING HOW CONTRAST IS REDUCED

Now that you have a better understanding of the reasons for using filters to reduce contrast, a more thorough explanation of how these filters actually work may be in order. Figure 6 provides a good explanation of what happens to the color negative when certain filters are used to reduce the overall contrast of a scene.

If you were to use a densitometer, which is a device that measures the optical density of the negative emulsion after it's been developed, and plot

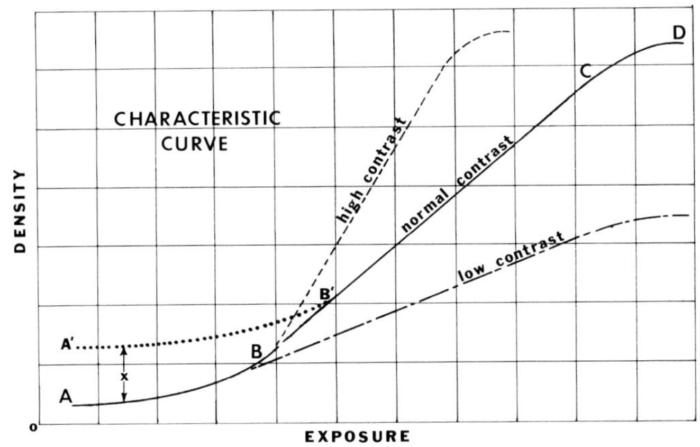

Figure 6 The characteristic curve of an emulsion. Section A to B represents the "toe" section, B to C represents the straight line portion, and section C to D is the "shoulder" of the curve.

those readings against similar increments of increased exposure, you would end up with a graph that demonstrates that for every specific increase of exposure, there is an increase in emulsion density.

The graph in Figure 6 is known as the densitometric curve of a negative emulsion, and it is usually referred to as the *characteristic curve* of that emulsion. An ideal emulsion has an equal increase in density for each equal increase in exposure. This results in a graph that has a straight line rising at 45° from the lower left corner to the upper right corner. The lower left corner represents zero brightness, or black, and the upper right corner represents the highest density possible, pure white, properly exposed. Any further increase in exposure would not create any further increase in density. Similarly, a black area with zero exposure would result in a clear negative base with no density at all. (The fog level—the extremely slight density produced by normal development—is purposely ignored here.)

Figure 6 shows that from section A to B, the density of the negative barely increases in relation to the increase in exposure. This section of the graph represents the extremely low light levels, which are the darkest shadow areas of the scene, and is the "toe" of the densitometric curve. Section B to C, the straight line portion of the curve, shows a relatively equal increase of negative density for each increase of exposure. Section C to D is known as the "shoulder" and represents the highlight area of the negative; additional exposure increases density only minimally.

The flatter the characteristic curve is, the lower the contrast of that particular emulsion. The manner in which the negative was developed can also determine the level of contrast. The upper line, steeply pitched, represents a high-contrast emulsion, such as a high-contrast stock used for titles or matte work. The normal contrast line can become slightly steeper when the negative is force-developed, thereby becoming more contrasty.

Section B to C represents the full printing range that can be reproduced in a properly timed print. On a normal print, A to B will reproduce as black, and C to D will reproduce as white. Although there is slight density in section A to B, it's below the printing range. If there were some way to increase the density of this low exposure area without affecting the balance of the image, it could be brought into the printable range. Evidently, it would take just a little more density to do that. Increasing the exposure doesn't help just the "toe" area because it would increase the densities overall and actually push some of them up into the "shoulder" area where they'd be lost. An increase in exposure would also necessitate an increase in the printing light intensity in order to print through the denser negative. This could leave you in exactly the same situation as far as the "toe" is concerned because the higher printer lights would record the lowest densities as black. All that is needed is a little additional density in section A to B without an increase in density from B to D.

Fog and low-contrast filters do just that. They add a bit of controlled whiteness to the darkest shadow areas of the image. The filter manufacturers don't claim that their filters bring out shadow detail. Instead, they claim that by sufficiently building up the shadow density, the contrast range of the scene is reduced and the low-light-level densities are brought into a printable range. In Figure 6, the dotted line from A′ to B′ represents what happens to the "toe" when a fog or low-contrast filter is used. The toe density is effectively increased by the degree indicated as X, which brings that area into the printable range and reduces the overall contrast, as indicated by the new line A′ to D. This new line is less steep than the original A to D line, indicating a lower contrast. The filter manufacturers also claim that using a fog or low-contrast filter effectively increases the negative density through the addition of controlled whiteness; therefore, you can decrease the exposure by about one-half stop and still get a normal-looking print with reduced contrast. By overexposing one-half stop, the filter creates a greater degree of color desaturation with normal printing.

Knowing just what the densitometric curve of a film emulsion is can enable you to evaluate the inherent contrast of that particular emulsion. Assuming that a 45° straight line portion of the curve represents a normal contrast, then a curve that has a smaller angle represents an emulsion of lower contrast. Compared to the increase in exposure, there is less of an increase in negative density. Conversely, a more steeply pitched curve indicates a more contrasty emulsion. For every increment of exposure, there's a greater increase in film density. When plotting the densitometric curve of a color emulsion, three lines must be considered: one each for the red, green, and blue layers. The manner in which the three emulsions of the same film base superimpose themselves on the curve indicates their response to exposure. The straight line portion of the graph should show very little or no variance in these lines.

If you understand the technical workings of contrast reduction, you will have another tool with which to control contrast and, therefore, the image.

SELECTING A FILTER TO USE

How do you decide which filter to use? For starters, assume that you are testing low-contrast filters to reduce the contrast of a scene. When you start your tests, view the scene through the filter, taking care that sunlight doesn't hit the filter and shield it from the light of an overcast sky. If light from anything other than the scene should accidentally strike the filter, front or back, you could have a problem. It could create an excessive effect or cause a flare on your picture. Always shield the filters in front of the lens with a matte box and an additional sun shade, if needed. (See Figure 3.)

For your test, try to be objective. Look at the shadow or dark areas, look at the bright areas, and look at the areas where they meet. Move the filter away from your eye while focusing on the scene, then move the filter back again. If you do this enough times, you'll begin to see what the filter is doing to the dark and light areas. When you understand that, put the filter in front of the lens and look through the camera on the ground glass.

The lens and the ground glass can strongly influence your judgment when choosing a filter. Some lenses are more contrasty than others, and some ground glasses can affect what you see. It isn't always a case of "what you see is what you get!" You must sometimes interpolate, taking into consideration the qualities of the lenses and the ground glass. When you become familiar with particular sets of lenses and use only one or two types of cameras, your job is simplified.

With a series of prints or slides to study, you'll see the different effects caused by the filters you've tested. You'll also learn how exposure can affect the results of using a particular filter. As an example, if you have purposely overexposed a scene by one-half stop while using a regular fog #3 to mute the colors, you may find that the lab has increased the printer light for normal density, thereby offsetting your desired look. You must learn how to give printing instructions to the lab so that the effect you want will not be lost in the printing process. Scrutinizing your tests or dailies with a lab technician can help ensure the final results.

Keep ample, precise notes, and always try to be objective when interpreting test results. A scrapbook can be very helpful. On all of my films, I have the camera assistant keep a notebook using a rubber stamp made as follows:

Date ______________	Scene __________
Day/Night __________	Int/Ext __________
Lens ______________	Height __________
Distance ___________	Filter __________
Meter Reading _______	Stop Used _______
Lab Notes ______________________	

This notebook serves as a permanent record of just which filters I used and whether I altered the exposure from the meter reading. As the production continues, it becomes a handy reference against which to check upcoming shots. It is especially helpful if a scene must be reshot; all of the pertinent information is there. It's true that the script supervisor usually keeps a record of some of this information, but not all.

Some experimenting can be done on the job as long as you don't overdo it! You can use a filter for a particular situation, and if the results are what you expected, you know you're on the right track. However, don't suddenly throw in a heavy filter just to see the effect! If it isn't correct, you may well ruin the scene!

Even experienced directors of photography make a series of tests before principal photography begins so they can determine what to do to obtain the look they want. In fact, these tests are an indication of experience. Some inexperienced cinematographers mistakenly believe that tests aren't important. Testing during the preparation of a film is not an expensive proposition, but once the production is underway, it can be very difficult to find the time for it.

Tiffen produces a kit of 50 different filters of various types: fogs, low-cons, diffusion filters, nets, and so on. The kit, which contains a one-inch round filter of each and a holder for viewing through, could be just the right teaching tool for individuals or classes. (See Figure 7.)

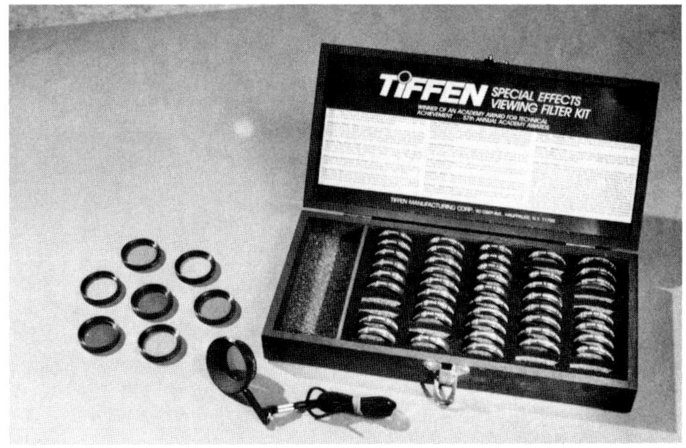

Figure 7 This Special Effects Viewing Filter Kit was assembled by Tiffen and contains an assortment of 50 different filters, a filter holder for viewing, and a brief description of the filters. The kit contains fogs, double fogs, low-cons, corals, diffusers, enhancing filters, stars, soft-contrast filters, and black, red, white, and skin tone nets.

This is a good time to start testing the basic filters discussed in this chapter. Remember:

Film stocks affect filter effects.
Printing density affects filter effects.
Development affects filter effects.
Lenses affect filter effects.
The ground glass affects the visual effect only.

The next chapter deals with more specialized types of filters. Working with this first group of filters will help you understand the more complex group.

In recent years Tiffen has added several series of filters that can also be used to reduce contrast; the Ultra Contrast Filter, the Pro Mist Filter, and the Soft Contrast Filter. Each of these series has subtle effects. The Ultra Contrast Filter will lower contrast: by slightly brightening the shadow areas without creating any flare or halo from light sources within the frame. The Pro Mist Filter will subtly reduce the contrast and add a bit of softening of the image, and in the heavier filters will slightly mute strong colors. The Soft Contrast Filter will reduce contrast within the scene without affecting the black areas, and it will tend to soften intense colors. The Tiffen Filter company offers technical data on all their filters. (See Appendix: Filter Manufacturers List.)

Chapter 5
Special Filters for Special Effects

As defined in Chapter 4, a filter is any device used in the optical system of a camera that modifies or changes the quality of the light passing through the lens to the sensitive emulsion. Under this definition, a diffuser is a filter too, and the effect created by a diffuser cannot be duplicated by a film lab. Diffusers are worth their weight in gold! They are the cinematographer's magic brush with which to retouch fine facial lines and transform an ordinary landscape into a magical fairyland.

DIFFUSERS

Diffusion is probably the least understood of all the filter effects. You may master the art of diffusion only after years of observing your results on the screen. Even then, because the accepted styles of cinematography often change, it may never be a perfected or agreed upon art. Diffusion should be a virtually invisible effect. At its best, the results are never apparent. Because the choice of diffusers is entirely in the hands of the cinematographer, the decisions made during filming are irrevocable. Labs can neither add nor take away the effect of camera diffusion as they can many other effects. The art comes in knowing exactly the degree of diffusion to add to a scene; too little and the effect is nil, too much and the effect is obvious, observable, and possibly a distraction. An additional challenge is to select the right amount of diffusion not only for a particular close-up but also for an entire sequence.

A diffused image should in no way be confused with an out-of-focus image. In an out-of-focus image, no part of the subject is sharp at the plane where the lens is focused, whereas in a diffused image, some lines are sharply represented and others are soft. A diffuser is designed to refract some light rays passing through the lens in such a way that they superimpose a slightly blurred image over a sharp image. These combined images create the diffused effect. The degree to which the resulting image is diffused is proportional to the percentage of refraction caused by the diffuser.

Many varieties of diffusion material are available, and each has a variety of features. The category of glass diffusers, for example, includes laminated diffusers in which a gelatinous material with the ability to slightly spread some of the light rays is mounted between two pieces of thin, optically clear glass. Molded glass diffusers with slightly raised ridges can bend only some of the light rays, thereby "spoiling" the otherwise perfect clarity of the glass and thus the sharp resolution of the lens. Some glass diffusers have an etched pattern that refracts a few of the light rays while letting most pass through unaffected. Other laminated glass diffusers contain a piece of netting material. There are many plastic diffusers and holders for plain unmounted netting. Any material that refracts, or bends, some light rays while permitting most to pass through unaffected can be used for diffusion.

Diffusion effects can be created from myriad materials. You can use a thin smear of petroleum jelly on a piece of optical glass to dramatically affect the lens image, or crumple a piece of plastic wrap, stretch it out flat, and observe the image through it.

The advantage of glass diffusers is that they are stable; their effects are always the same and, with experience, are predictable. They usually come in sets of five, marked #1 through #5 or A through E; each giving a progressively stronger effect. If you make your own diffuser by mounting some material between two pieces of glass, know that ordinary picture glass is not the best choice. Often it's neither optically clear nor completely colorless. It's best to use optically clear, water-white glass.

In the early days of black-and-white filming, before the advent of diffusion, cinematographers had a favorite lens for close-ups. It was approximately 100mm, of questionable resolution, and perhaps contained several optical aberrations, but it rendered a lovely soft-focus image flattering to actors. No other diffusion was necessary! As lenses were perfected, with fewer and fewer aberrations, and the resolution of the film stock increased, a means had to be found to make the sharper image more flattering to older faces. Cinematographers supposedly used a bit of spittle on the lens itself; it diffused the image and was easier to clean off than petroleum jelly!

The ratio of soft image to sharp image determines the degree of diffusion in a particular filter. Some diffusers work two ways at the same time. In addition to refracting some of the light rays to produce a softened image, some diffusers also tend to flare some of the highlights throughout the image. This mutes the colors and grays the blacks somewhat, which reduces the overall contrast and helps create a more flattering close-up.

Diffusers of different manufacturers vary in the ways they work. Some reduce the contrast markedly as they soften the image, others reduce the contrast only moderately, and some do not affect the contrast at all. The H&H regular diffusion filter reduces contrast substantially as it diffuses, and that combination can be very flattering. (See Color Plates 18 and 19.)

H&H has recently marketed a black dot texture screen diffusion filter that does not reduce the contrast or add a halo around very bright areas, but it does soften the image. The black dots in this diffuser reduce light transmission; thus exposure must be increased slightly to compensate, and the more open stop helps reduce sharpness. Other glass diffusers generally don't require any exposure change. Pancro-Mirror's diffusers are similar to the old Mitchell-type diffusers: they diffuse wonderfully and only barely affect the contrast. Tiffen produces both acrylic diffusers and nets laminated in glass. The black net diffuses without affecting the contrast. The white net, however, affects both sharpness and contrast. Some nets require about a one-quarter stop increase in exposure.

Before the days of glass diffusion (and before my time in the industry), if a director of photography had to eradicate harsh facial lines, he put a silk stocking over the lens! The very fine mesh of the silk stocking, when tightly stretched, had the wonderful effect of softening the image and the contrast while blending facial blotches to the point where they were no longer discernible.

Today, cinematographers comb fabric houses, gathering bits of netting materials that vary in type and thickness of thread, size of the weave, and color of the thread. Because of its shiny surface, nylon thread softens more than cotton thread, and the closer the weave, the greater the softening effect. Colored net diffusers affect the image very differently than glass diffusers. Nets of black thread do not affect the contrast or the color. Nets of white thread soften the image and reduce the contrast. If the net material is a color, it will impart a slight hue over the image. If the color of the thread is a warm hue, it can be very becoming to flesh tones. Tiffen manufactures a product called Softnet, which has various weaves of netting—colored and black and white—laminated between two pieces of optical glass. The glass protects the net and keeps it ready for consistent use. Directors of photography who want to make their own net diffusers can use special frames on which the net material can be stretched and held in place with spray adhesive along the edges. The net can also be mounted between two pieces of optical glass.

Most glass diffusers do not require an increase in exposure, but net diffusers do require a slight exposure compensation because the threads prevent some of the light from reaching the film. The exposure factor of a net can be anything from one-eighth to one-half stop, and it can be checked with a meter.

Any glass placed in front of a lens with a long focal length usually distorts the image so badly that it is unusable. If diffusion is required with long lenses, a net works more effectively than a glass diffuser. White nets of various densities can match a long lens to a normal prime lens used with a fog or low-contrast filter. Wide-mesh white net reduces the contrast of the image and affects the sharpness only minimally. Because long lenses (500mm or more) cannot tolerate glass filters, they usually have gelatin filter slots in their lens barrels. These slots are in an optimum optical position to avoid distortion or focus shift. Gelatin filters can be used when matching or balancing dictates the choice of a filter.

Nets placed in front of a lens can also cause problems! Depending on the focal length of the lens and the stop being used, a net can sometimes become slightly in focus and appear as a soft pattern over the image. The probability of this happening is increased as the focal length decreases and the F/stop increases. To avoid this problem, cinematographers sometimes mount the net on the rear element of the lens. The manner in which the image is affected by the net varies depending on its position in front of or behind the lens; testing will confirm the best position for your purpose. (See Color Plate 25.)

Net filters were once used to create a vignette effect. The center of the net was removed while it was stretched on a net frame or over the front of a matte box. In the days of black and white, cinematographers used a lighted cigarette to burn out the center of the net as they watched their efforts on the ground glass of the camera. I guess it was simpler than using scissors! Whatever the method, if the remaining net material was black, it diffused and darkened only the edges of the frame. This effect was used in long shots as well as close-ups. In the close-ups, perhaps only the eyes were filmed through the clear area, leaving the balance of the frame diffused and darkened.

If used similarly, a white net created a vignette of diffused edges but, instead of being darker, the outer edges contained muted colors due to the light-dispersing action of the white net. This effect was used for creating fantasy landscapes and dream sequences, and for close-ups. Another vignette effect was created by smearing the edges of optical glass lightly with petroleum jelly, leaving a clear glass oval in the center. White toothpaste used in the same way produced a vignette with diffused, white, foggy edges. Whether using petroleum jelly or toothpaste, only a very thin film was necessary.

When and How to Use Diffusers

The most prevalent use of diffusers is to flatter by softening facial "realities" to the degree that they become unobtrusive. Diffusion does not work by itself, however. To use diffusion most effectively, cinematographers must first light the subject in the most becoming manner. The harsh light from a fresnel spotlight causes sharper facial shadows than the light from a reflected source such as a soft or bounce light. The strength of the diffuser depends mainly on the size of the image. Generally, the larger the close-up, the heavier the diffusion; the wider the shot—head and shoulders, waist shot, full figure—the lighter the diffusion.

You must select the diffusion effect desired by viewing the ground glass at approximately the lens stop that you will use. An excellent way of making this decision is to look at the subject's eyes and select the diffuser that maintains a sufficient degree of sharpness in the eyes. You may find that the ground glass image looks slightly softer than the projected screen image. A small, sharp highlight often heightens the sharpness of the eyes even when a heavy diffuser is used. The eyes are a good place to decide about diffusion because that is where the audience mostly looks.

When filming a sequence, directors usually start by shooting the master, or long shot, first, then they move the camera in progressively closer until the last setup is an extreme close-up. If you are aware of the planned coverage, you can select in advance the different degrees of diffusion to be used in the different shots. That is, you would start with the least diffusion in the widest shot and use increasingly heavier diffusion in the closer shots. When the various shots are edited together, the more heavily diffused close-up will match nicely with the preceding shots. If no diffusion is used in the wider shots, cutting to the diffused close-up can be jarring.

What if the director wants to start with a long shot that requires no diffusion and then dolly into a close-up of a character's face, which does require diffusion? Variable diffusers can solve this problem. They consist of two oblong diffusers, each varying from clear at one end to medium diffusion at the other. The diffusers are mounted in opposite directions in a sliding track that allows them to be moved to overlap each other. In the long shot, both clear ends of the diffusers overlap in front of the lens. As the camera moves closer to the actor or the actor moves toward the camera, the glasses are moved (either manually with a hand knob or by motor) so that the diffused ends begin to overlap, adding their effects together to create heavier diffusion. The control knobs are indexed so that any degree of diffusion can be obtained, and the whole procedure can be easily repeated for additional takes of the same scene. Originally, Mitchell produced the only variable diffusers, but now they can be rented from Panavision and other rental houses. Because the variable diffuser is not ordinarily carried throughout the production, you must plan ahead for its use.

When filmed in an extreme close-up with today's ultrasharp lenses, even the most beautiful and flawless face can show exaggerated skin texture. The pores of an actor's skin can be downright distracting. Diffusion helps blend the facial tones so the audience can concentrate on the actor's eyes and the words and the feeling of the scene.

Often in the past, and still today, it was generally the practice to use diffusion for women's close-ups, but not for men's. This double standard is seen as sexist today; why should a woman's appearance require "help" or "softening" when a man's does not? That discussion aside, when male and female close-ups will be intercut, it's best to use some diffusion on the male—usually one or two degrees less than that used on the female when the image is the same size. Why not use the same degree of diffusion? Women traditionally wear more make-up in front of the camera than men do. The small degree of difference in diffusion will make the two close-ups look the same. There is no formula for determining the degree of diffusion relative to the size of the close-up; each shot must be decided individually.

What if there are two large close-ups within the same frame, that is, a tight two-shot of a man and a woman? H&H and Pancro-Mirror both make sets of half-and-half diffusers; one side is clear, and the other side has a degree of diffusion ranging from D½ to D7 or A to E. These diffusers allow one person to be diffused and the other person, in the clear area, to be kept sharp. If the two actors require different degrees of diffusion, a full diffuser can be added to a half-and-half diffuser to create heavier diffusion on one side of the frame and lighter diffusion on the other. You can special-order any combination of these filters. The half-and-half diffuser is also used when a person in the foreground needs some diffusion while the background scene doesn't.

Diffusion works like an airbrush, but it isn't limited to giving beauty treatments to close-up faces! It's a lot of fun to plan the use of various techniques within a given film so that the sequences don't all have an identical look. For example, night exteriors can be filmed without any filters, day exteriors enhanced through low-contrast filters, and night interior love scenes made more romantic through an effective combination of a fog, a coral, and a bit of diffusion! Landscapes can have a fantasy quality if properly diffused, and the colors can be muted slightly with double fog filters. The almost limitless combinations make films more interesting to experience and more beautiful to behold.

Although most filters affect 35mm and 16mm film in the same ways, such is not the case with diffusion filters. A diffusion filter will have a stronger effect on 16mm than on 35mm. As a general rule, try at least one degree less diffusion when shooting 16mm film; that is, Mitchell B or an H&H D-2 in place of a Mitchell C or an H&H D-3. Make additional tests if you switch from 35mm to 16mm film or vice versa. It's also a good idea to retest the effects of the fogs and low-cons with 16mm equipment if you're used to working in 35mm. Filming specifically for television may also impose some restrictions on filters. Because home television receivers do not have the resolution of theater screens, it might be prudent to use slightly lighter filtration in most cases—except perhaps fantasy sequences. These guidelines may change when high-definition television is adopted.

When filming trick shots where thin wire is required (to fly an object or an actor), a bit of diffusion can save the day by making the wire less visible. This type of shot is also greatly aided by hiding the wire against a variegated background or by using a bit of camouflage paint on the wire. However, the scene must first be judiciously lit before any of these tricks, including diffusion, will work.

A screenplay that calls for a special look—for example, a land of make-believe, a flight in outer space, the fantasy of the inner mind, or just a special sequence in an otherwise normal film—gives the cinematographer a perfect opportunity to call into play many combinations of effects. These can include lighting, filters, and special laboratory techniques.

Testing procedures are especially important when diffusion is used. Decisions made at the time of testing often change as the filming progresses because the cinematographer is constantly educated by the dailies. Tests should be thought of as guideposts or jumping-off places. The director of photography is probably the only person who's fully aware of the filter effects when the dailies are shown. There's certainly nothing wrong with making minor improvements the next day at work and continuing to do so until the desired look is achieved. Just make sure that your day-to-day changes are subtle so that when the scenes are edited, you won't be embarrassed by glaring mismatches!

STAR FILTERS

The star filter is an image controller that has been much used and much abused. It creates a star burst out of any point source of light, including candles, street lights, electric bulbs, theatrical spotlights, reflections from the sun, jewelry, and sequins. The star filter is made of clear glass with etched lines. If the lines are at right angles to each other and are engraved or etched 2mm to 6mm apart, the filter creates a four-point star. Adding one more line so that all are 60° apart creates a six-point star; a filter with four lines that intersect at equal angles creates an eight-point star. The effect of the star filter can be fascinating when appropriately employed. The engraved lines intercept a bright light source and spread the beam of that light along the engraved lines, creating the star effect. The light beam elongates in proportion to the intensity of the light source. As a side effect, the star filter lends a bit of softness to the scene, almost like a light diffusion glass, and this can be used as an effect in itself, even if there is no bright point of light to create a star. (See Color Plate 26.)

In the event that multiple cameras are used and there isn't a star filter for each camera or a star effect is desired but there's no star filter available, the cinematographer can create a star effect very simply. On an optical flat, a glass that is free of any color or optical aberrations, make some very light lines of petroleum jelly with a cotton swab in the area you want the star. This handmade, on-the-spot filter is very versatile because the petroleum jelly star can be placed anywhere on the glass, allowing the effect to appear only in the section of the frame where you want it. It takes a bit of experimenting to get just the right amount of grease on the glass: if it's too heavy, it will diffuse the image too much; if it's too light, the star effect will be weak. The main disadvantage of the handmade star filter is that petroleum jelly is not stable. It can smear and will not necessarily keep from one day to the next, but once the technique is learned, it's simple to duplicate.

Another substitute for a star filter is a piece of metal screening, such as a section of window screen, that's been buffed to a high polish. Used as a filter, it spreads a point of light into a star. Some cinematographers have pieces of chrome-plated screening that work excellently. All metal materials require a slight exposure compensation.

With any type of star filter, the cinematographer must study the effect on the ground glass before filming. Star filters are generally round so that the points of the stars can be turned to whatever angle is best for the composition. If the lines of the four-pointed star are set exactly vertical and horizontal, a cross is created, which may be fine for some scenes but not for others. Therefore, look at the angle of the star before committing it to film. If the star filter is mounted directly on a lens that rotates when the focus is shifted, the star effect will also rotate during the shot! That could be a dizzying disaster!

DIOPTERS

A diopter is a lens—a simple, one-element glass that, when placed in front of the filming lens, enables the lens to focus at a closer distance than that for which it was calibrated. The diopter is like a magnifying glass that enables the human eye to focus on extremely close objects. When using a zoom lens for extreme close-up work, a diopter must often be used because many zooms cannot be focused closer than three or four feet. With the correct diopter, a zoom can be focused as close as two feet, perhaps closer, and still have the full range of its zooming capability. Diopters come in different strengths; including +¼, +½, +¾, +1, +1 ½, and +2. The higher the number, the stronger the power. Diopters require no exposure compensation unless you are filming at less than 10 times the focal length of the lens. At these closer distances, lens formulas change the effective F/stop. There is a formula for figuring the effective F/stop when the normal markings no longer hold true. [See the *American Cinematographer Manual*, Sixth Edition, edited by Fred H. Detmers (Hollywood: The A.S.C. Press, 1986: pp. 375–379) for extreme close-up lens formulas.] Full diopters can also be used on prime lenses when they are required to focus beyond their calibrations.

Split Diopters

A split diopter lens is a semicircular diopter that allows part of the filming lens to focus on a close subject while the other half of the lens (not covered by the diopter) focuses on a distant subject. The split diopter is comparable to "Ben Franklin" eyeglasses that allow you to focus on a close subject through the half lens below and then focus on a far subject by looking up and over the half lens. These half glasses, now widely sold in drugstores, are graded according to their diopter power. The semicircular split diopters are usually mounted in a special holder on the matte box that enables the cinematographer to rotate it to any angle and to slide it to any position to accommodate the foreground subject within the frame. (See Figure 8.)

The practical application of the split diopter is apparent when you consider that, even with a wide-angle lens, a small lens stop is required to carry the focus from a foreground subject, as close as 18 inches, to a background subject perhaps 20 feet away. Working indoors at relatively low light levels necessitates a great change in lighting to accommodate the required lens stop for the depth of field. The split diopter can do the job with the lens wide open if need be. The filming lens focuses on the far subject, and a split diopter of proper strength is positioned to bring the foreground subject into sharp focus. (See Figure 9.)

The selection of a split diopter depends on the focal length of the filming lens and its distance to the close and far subjects. There are charts calculated to help select the proper diopter strength; but it's relatively easy to use a hit-or-miss method by trying first one and then another of the diopters. Split diopters are generally rented in sets of six different strengths, along with a special holder that mounts on the matte box.

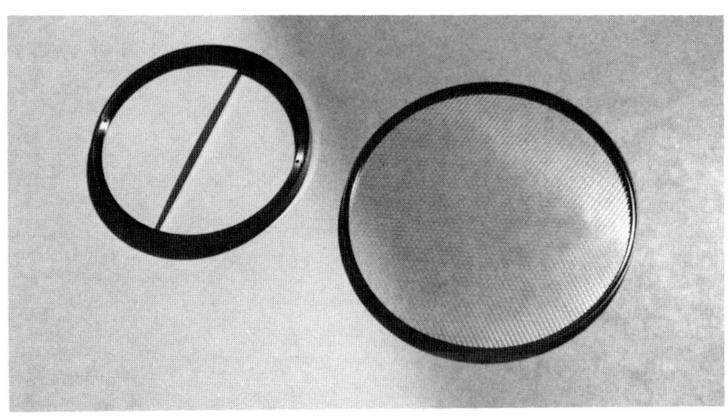

Figure 8 The filter on the left is a split diopter made by Tiffen. The larger filter is a star filter.

A few cautions are in order regarding the split diopter. Extreme care must be taken to prevent the line of demarcation—the edge of the split diopter—from showing. The edge becomes visible if it falls across an object, throwing part of that object out of focus. The edge is more easily concealed in the lower power split diopters than in the stronger ones. The line can often be hidden by aligning the edge with a door frame, a hanging drape, a shrub, or a plain wall, but the line is easily seen if any object overlaps it. You must also make sure that nothing moves through the edge of the split diopter.

If you change the focus of the prime lens during filming, it will affect the close focus through the split diopter. Shots using a split diopter must be set up very carefully. One common use is filming the driver and the front seat passenger with a camera mounted on the door of the car. In daylight with a small lens stop, you probably won't need a split diopter, but if it's a night exterior with available street light in the background, it's not feasible to build up the light level on the subjects in order to stop down the lens for depth. In this case, the split diopter does the job perfectly at a wide open F/stop as long as the actors do not lean past the invisible edge. The actor in the foreground could lose a nose, or the background actor an ear !

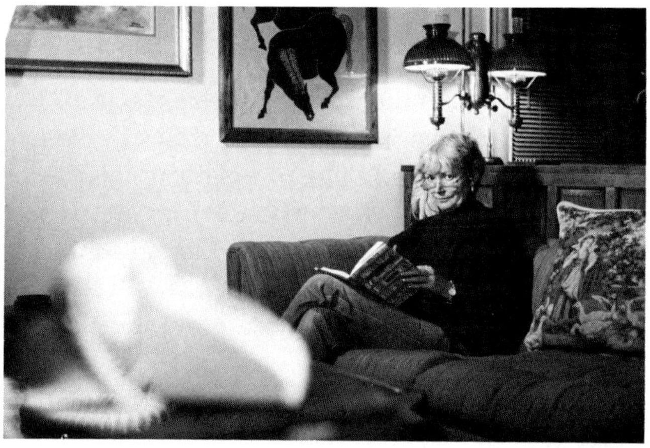

9A

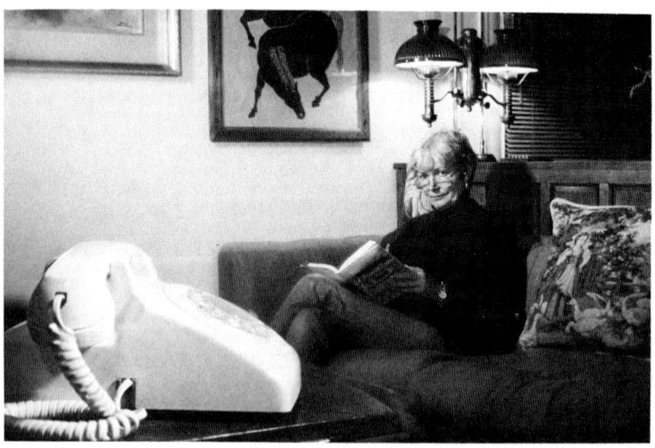

9B

Figure 9 Figure **9A** was shot with a 35mm lens at F/2.3. The focus is on the actor; the foreground phone is blurred. The same lens, same stop, and same focus were used for Figure **9B**. A #1 split diopter was placed in the lower left corner at an angle and positioned to cover just the telephone.

There are a few tricks that make split diopters even more versatile and fun to use. Suppose there's an actor in the foreground, camera left, facing the camera and another in the background, camera right. The split diopter brings them both into sharp focus, but then the background actor walks forward as the foreground actor turns his back to the camera and moves slightly, causing the camera to pan. While the camera pans, the assistant cameraperson can unobtrusively slide the split diopter out of the frame, leaving the lens clear for any further action and focus changes. Panavision's holder has slides with stops that can be preset to position the split diopter to be moved in or out of the frame in sync with the action. Two split diopters can also be used together if set at angles to one another so that there is some clear space between them. The actor farthest from the camera occupies the clear area with the focus set at that position. Two other subjects are then positioned, each behind one split diopter, at distances that hold them in sharp focus.

The split diopter can be used effectively in any area of the frame—top, bottom, or side—or at any angle. It's a wonderful device to keep in mind when planning the best way to shoot a scene or to use when presented with an unusual situation that can take advantage of its dramatic effect.

SPECIALTY FILTERS

Filters offer such a wide range of image enhancements that filter manufacturers have created a rainbow of colored filters and a variety of other image modifiers. Almost any filter that a cinematographer dreams up for a special look can be made to order. Harrison & Harrison, Tiffen, Lee, and others are most obliging in this respect. These manufacturers have made graduated colored filters that combine two colors in the same filter. When a cinematographer wanted a filter to create a warm glow along the horizon, H&H devised a filter with a horizontal orange strip about three-quarters of an inch wide that went right through the middle of an otherwise clear glass. This strip filter could be raised or lowered as needed, and the cinematographer had his effect!

Harrison & Harrison recently carried the strip filter one step further with the new and beautiful sunset filter. In a graduated filter, they combined a strip of coral #8 with a deep cyan for the top, leaving the bottom half clear. Toward the end of a day with an otherwise bright sky, this filter creates a great sunset effect. As with many filters, you should try different exposures for the optimum effect, and always be aware that the lens stop not make the filter too sharp due to a small F/stop and short focal length lens. Select an F/stop that gives you the desired effect without delineating the filter, then control the exposure with neutral density filters, or by altering the shutter speed for still or video cameras.

Another new H&H filter, the first of its kind, is the "misty morn" filter. This strip filter uses fog in the middle and has clear filter areas above and below. In the hilly area where we live, I've seen this effect created naturally, and the filter replicates nature exactly. When positioning the effect to a scene that will be photographed, control the misty fog strip through the lens stop. The larger the aperture opening, the wider the strip effect. The smaller the aperture opening, the narrower the strip effect. Correct exposure is obtained with neutral density filters or shutter speed after the width of the strip has been determined.

When working with any graduated filters, be extremely cautious about camera movement. Tilting the camera will immediately destroy the filter effect. Some panning can be done if carefully watched. Of course, no action can cross from the clear area into the filter area.

Eastman Kodak is probably the largest producer of special filters for experimentation. They create filters for many scientific purposes, from ultraviolet wavelengths to the infrared end of the light spectrum. They manufacture practically all of the gelatin filters that are used in the film industry for both stills and motion pictures, including all of the light-balancing, color-correcting and color-balancing filters. Eastman also supplies many special filters for color lithography.

New entries in the filter industry have arrived from France. These are the filters produced by Cokin. The company was founded when photographer Jean Coquin wanted imaginative filters for his own use; he ultimately produced the Cokin Creative Filter System. Although originally designed for still photography, this system can be adapted to motion picture work as long as you stay within the confines of the available filter sizes. Cokin filters are made of plastic and are relatively inexpensive. An assortment of adapters fits them to all still camera lenses and many 16mm motion picture lenses, providing a simple, inexpensive way to experiment with various effects. Once you find the filter combination you want, any of the filter manufacturers can duplicate it in the more durable medium of glass.

As an indication of the virtually endless variety of filters available from Cokin and others, here are some descriptions:

- dream filters, which create diffused highlights, halos, and soft-focus images
- warm-tone filters, which are similar to coral filters
- multi-image filters
- blue filters
- sunset filters
- neutral and colored polarizers
- varicolored filters, which create some attractive and some bizarre images
- super-speed filters, which give stills the effect of an image smear associated with moving shots
- diffractors, which create a color burst from a point source of light
- colored graduated filters and diffusers

colored and neutral center spot filters, which are used for a variety of vignette effects

The Appendix contains a list of filter manufacturers and their entire range of filters and lens attachments.

THE LIGHTFLEX AND VARICON

Several years ago, an item was introduced to the film industry that is one of the most versatile attachments used in front of a lens. This ultimate image control device is the Lightflex. Designed by British director of photography Gerry Turpin, B.S.C., for his own use, the device was eventually purchased by Arriflex Corporation, which now builds and distributes it. The Lightflex resembles an oversized matte box and contains a curved glass reflector, and an overhead lightbox with a filter holder. The Lightflex is mounted on support rods in front of the lens, which views the scene through the curved reflector and also "sees" any reflection from the overhead light box. This overlays a controlled amount of light on the scene being photographed. (See Figure 10.)

The Lightflex is used three ways. First, it can be used to supply the fill light to the dark areas of a scene that are so thin in density that they print almost black and are devoid of detail. Although recorded on the negative, these dark areas fall below the printing range and are thus lost. Through its internal lightbox, the Lightflex adds enough exposure to those very low densities to bring them within the printing range of the negative. The added light is of low enough intensity not to affect the midrange or full density areas of the negative. The amount of reflected light that the Lightflex imposes on the film is controlled by a sensitive digital voltage regulator. The first use, then, is to increase shadow detail to within the printable range and thereby save extensive lighting of the shadow areas.

The second use is contrast control. The degree to which the Lightflex affects the image is controlled by an electronic dimmer with a digital one-tenth volt line indicator that adjusts the brightness of three quartz bulbs in the lightbox and uses a specially designed high diffusion glass to create an even, overall light source. The white light from the lightbox is reflected from the curved glass to the lens, and it can be increased to add light to the dark and shadowy areas of the scene. Because it does this without affecting the more fully exposed areas, it reduces the tonal range and thus the contrast. (See "Understanding How Contrast Is Reduced" in Chapter 4.) The Lightflex works as a variable low-contrast filter! On a shot-by-shot basis, it can easily be set to control and maintain contrast, and it can even be varied during filming. If the camera pans from a cross front light that requires a heavy low-con effect to a back light that needs less, the Lightflex's variable controller regulates the amount of light and adjusts the contrast. Something no low-contrast filter can do! The internal lamps that regulate the amount of white light cover a large brightness range without significant variation in color temperature.

The third use of the Lightflex involves adding a color to the internal filter tray. The light passing through the filter tray and reflected by the curved reflector overlays that color on the scene being filmed. It's not difficult to envision the unlimited range of the Lightflex in this regard. The colored gels in the filter tray are readily available gels or acetate theatrical colors. Some of the most popular colored filter combinations prepared by Lightflex

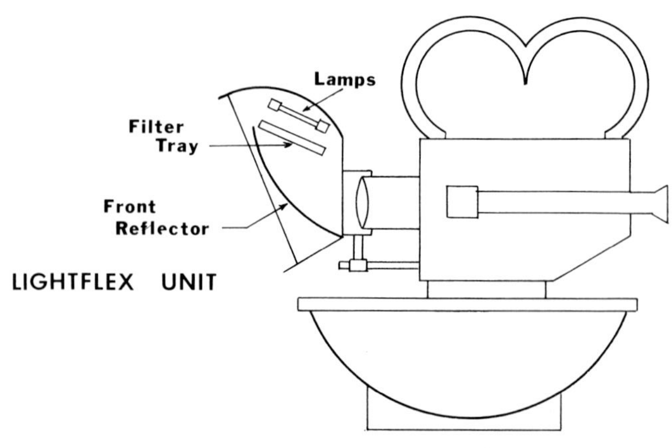

Figure 10 The Lightflex.

DECEPTIONS OF THE EYE (see page 3)

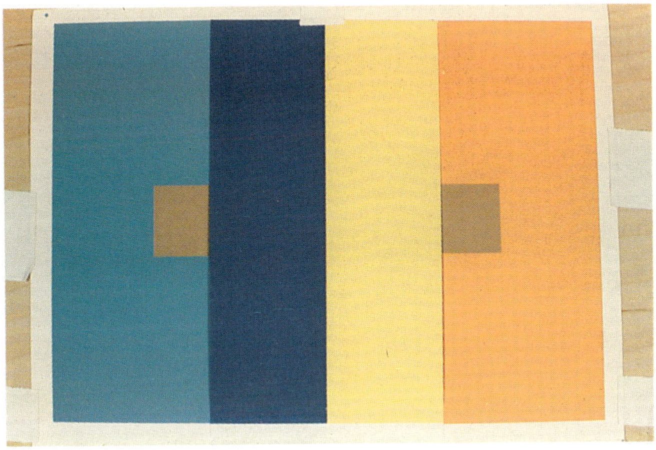

A

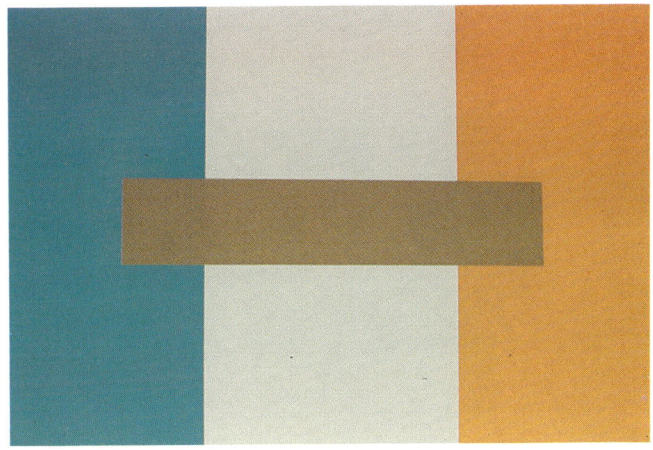

B

C

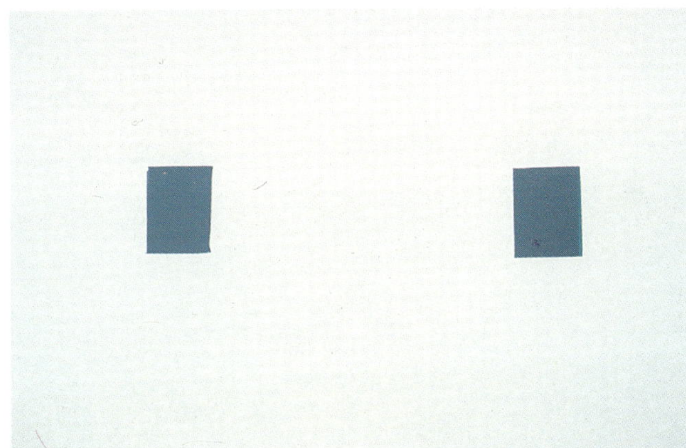

D

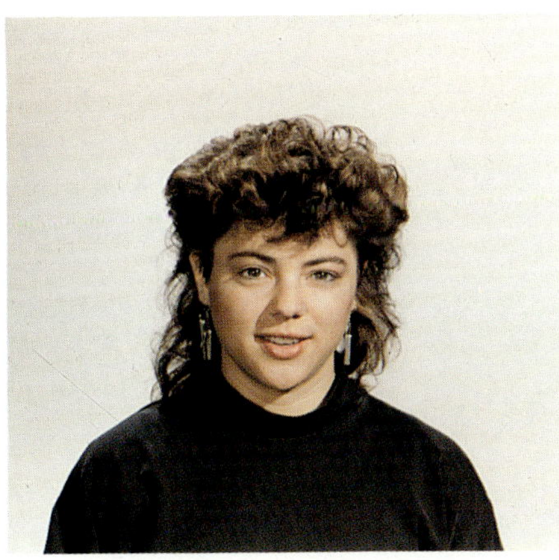

E

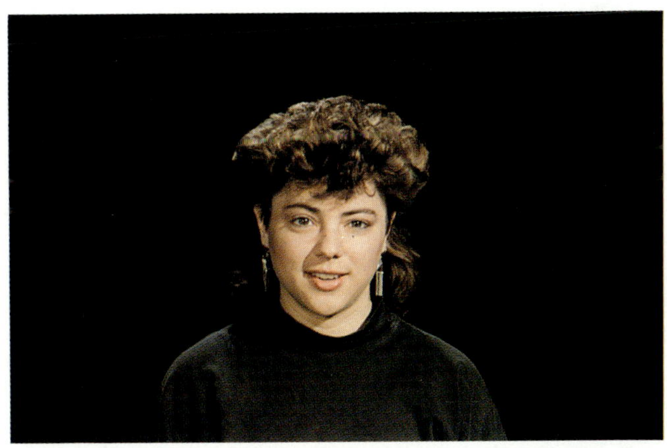

F

Color Plate 1 The brown square of color in the blue panel appears to be much lighter than the brown square in the orange panel. On the original color plate, the yellow and blue stripes can be lifted to reveal that the two squares are actually the ends of the same continuous brown stripe, as shown in Plate **1B**. The interaction of color relates to all colors and their adjacent backgrounds. The two green squares in Plates **1C** and **1D** are exactly the same color. Surround the center squares with white to see this demonstrated.

Study the young woman in **IE** and **1F.** Her hair, face, and sweater appear lighter against the black background than against the white. By surrounding her face with an opening in a neutral covering, thus illuminating the background, you'll see that the brightness of the woman is exactly the same in both situations. It may not be very noticable in these small illustrations, but the effect of the background brightness is heightened in the format of the large theater screen.

EXPOSURE CONTROLS THE IMAGE (see page 13)

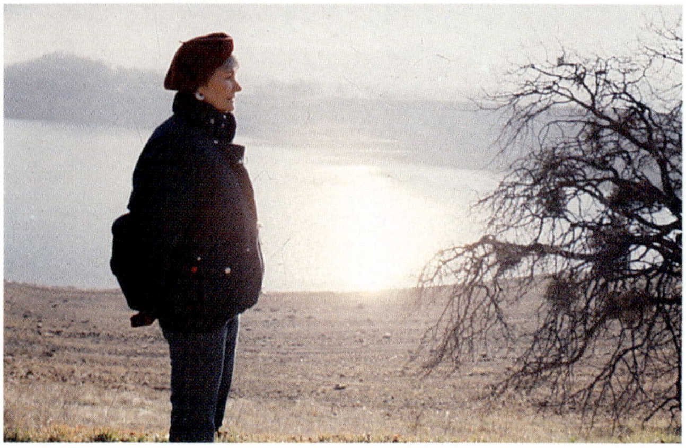

2A

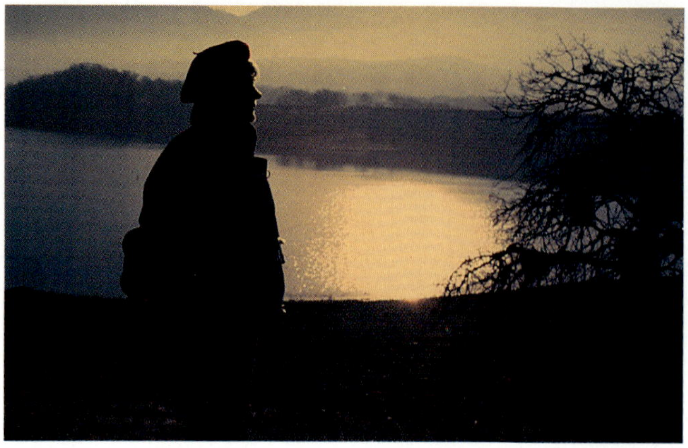

2B

Color Plate 2 Exposure in Plate **2A** was made for the face in this back light situation. As Plate **2B** shows, by reducing the exposure three stops, an entirely different image and mood are created. Exposure control affected the look of this scene.

LENS AND LENS STOP CONTROL THE IMAGE (see page 15)

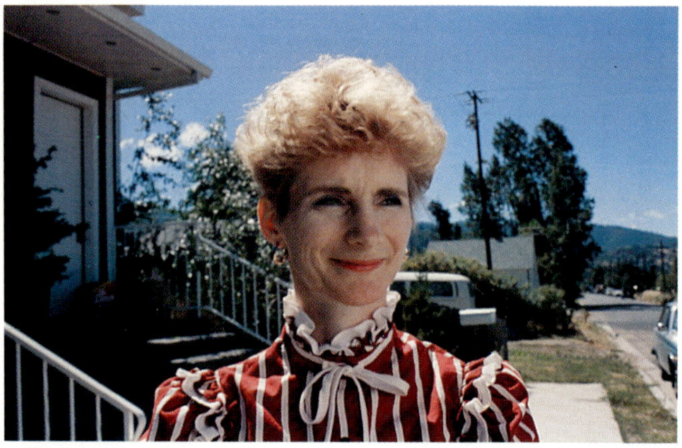

3A

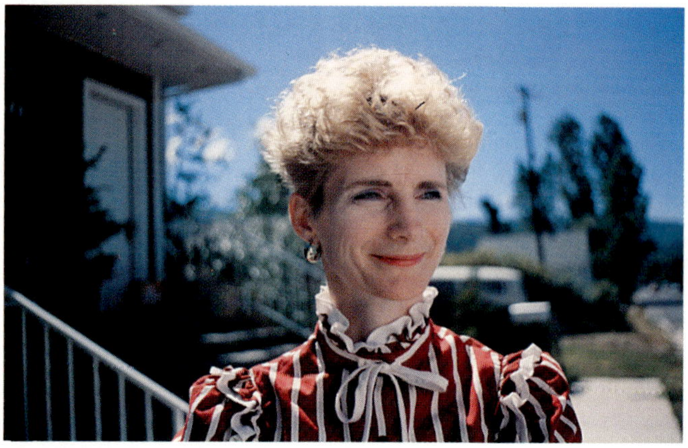

3B

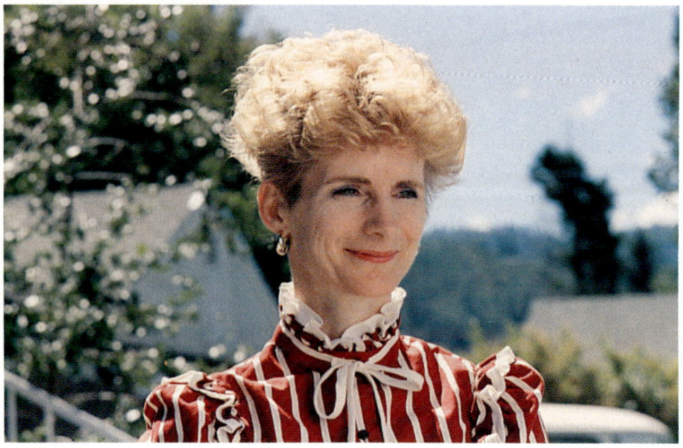

3C

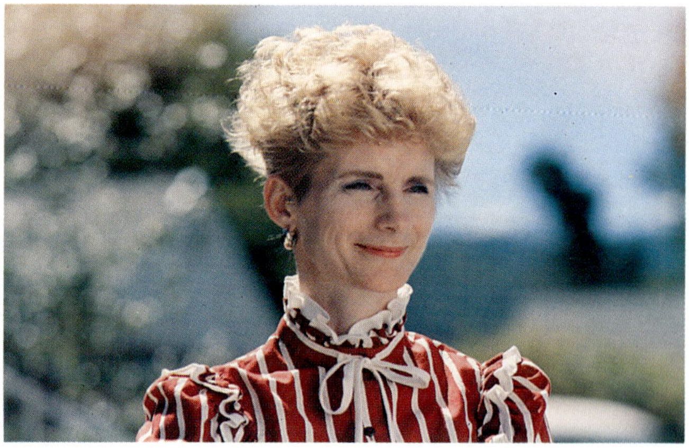

3D

Color Plate 3 Compare the scene in Plate **3A**, shot with a 35mm lens at F/11, to Plate **3B**, shot with the same lens at F/2.8 and a 1.2 neutral density filter to compensate for the exposure opening of four stops. Note that opening the lens stop gives the foreground subject a greater separation from the background which is now less sharp. Compare Plate **3C**, shot with a 100mm lens at F/11, to Plate **3D**, shot with the same lens at F/4 and a .90 neutral density (ND) filter. The 35mm shot and the 100mm shot have the same figure size. However, the 100mm lens eliminated the house and the telephone pole, decreased the background sharpness even more, increasing the three-dimensional effect and reducing obtrusive background objects.

CREATE IMAGE MOOD WITH CAMERA FILTERS (see page 19)

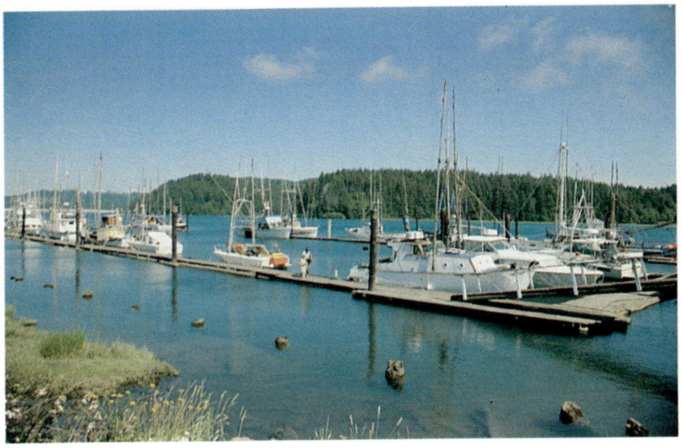

4A

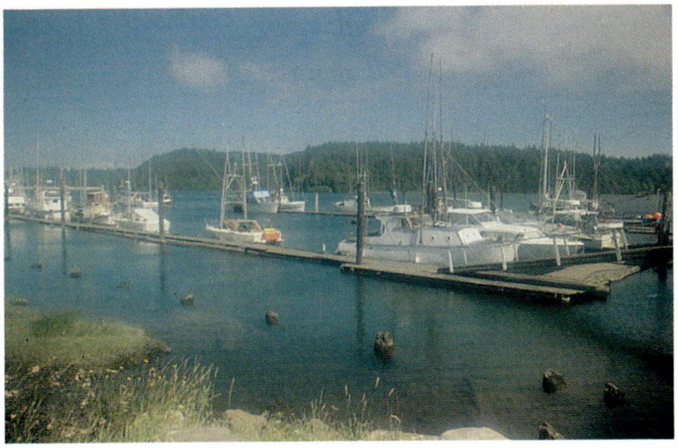

4B

Color Plate 4 The same waterfront scene was shot once without a filter and again with a regular fog filter #2 to make it appear more lyrical. Closely study the subtle differences between these two frames. In a practical application, the unfiltered shot (Plate **4A**) might be used for an adventure film, and the filtered shot (Plate **4B**) might be used for a fantasy or love story.

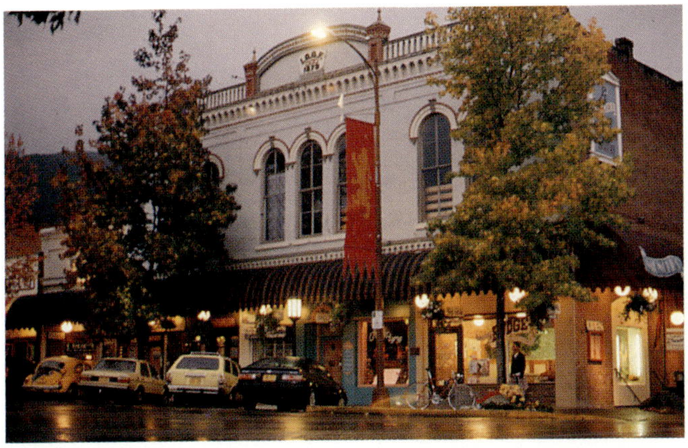

5A

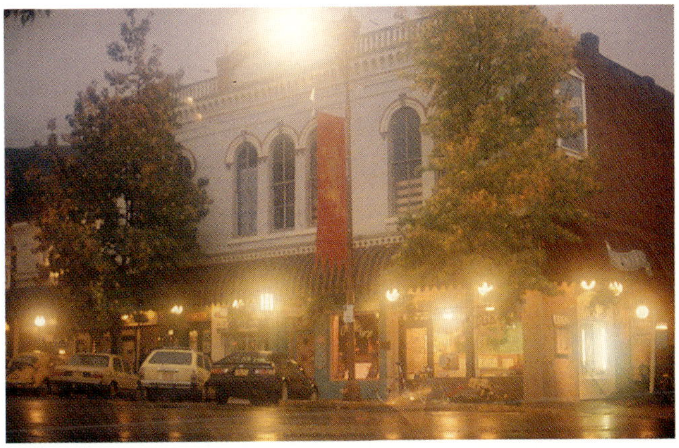

5B

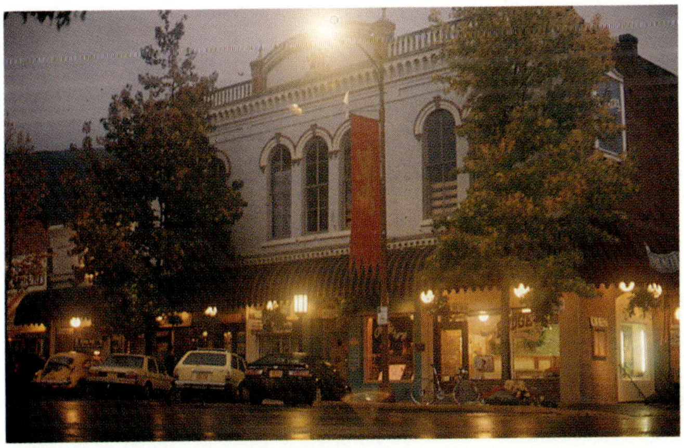

5C

Color Plate 5 No filter was used to shoot the street scene at dusk in Plate **5A**. Plate **5B** shows the same scene filmed through a regular fog #3. Not only is the contrast reduced and the resolution slightly softened, but the colors are also muted, and there's a halo around the light sources. This effect, typical of regular fog filters, can be lovely for scenes that require a special look. The same scene, filmed with a double fog #1, is shown in Plate **5c**. It too has reduced contrast and muted colors, but the resolution has not been softened and the halo around the bright areas has practically disappeared.

REDUCE CONTRAST WITH FOG AND LOW FILTERS (see page 20)

Color Plate 6 These scenes show how filters and exposure can control the contrast of a scene. This location was chosen because of the high color contrast, dark green walls, ivory trim, and white chairs. Note that as the strength of the filter increases, the effect of reduced contrast also increases. Using a .60ND filter permits a wider F/stop, which further reduces contrast. Also, as the exposure increases, the filter effect also increases. By using the filters and underexposing one-half stop, the density closely matches the unfiltered scene that was exposed normally because low-contrast and fog filters give the effect of additional exposure. Not only have the filters reduced the contrast, but they have also noticeably improved the green color of the wall by adding a bit of white light.

6A No filter, normal exposure.
6B No filter, normal exposure.
6C No filter, normal exposure.
6D Low-contrast #3 and .60ND filters, normal exposure.
6E Double fog #1 and .60ND filters, normal exposure.
6F Double fog #2 and .60ND filters, normal exposure.
6G Low-contrast #3 and .60ND filters, one-half stop over exposure.
6H Double fog #1 and .60ND filters, one-half stop over exposure.
6I Double fog #2 and .60ND filters, one-half stop over exposure.
6J Low-contrast #3 and .60ND filters, one-half stop under exposure.
6K Double fog #1 and .60ND filters, one-half stop under exposure.
6L Double fog #2 and .60ND filters, one-half stop under exposure.

An additional inadvertent effect can be seen here under close scrutiny. In the filtered scenes only, a ghost image of the white chair can be seen against the far dark green wall. This results from the use of two separate glass filters. The ghost image is the reflection of the chair from the front surface of the second filter to the back surface of the front filter, and then back into the lens! This happens only in very high contrast situations, and it is commonly seen when filming a candle flame against a dark background. It can be eliminated by taping the two filters together so that there's no air space between them or by using a matte box with a tilting filter holder.

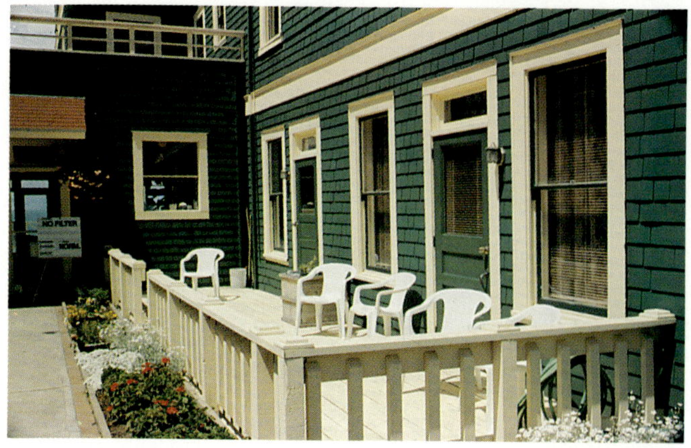

6A

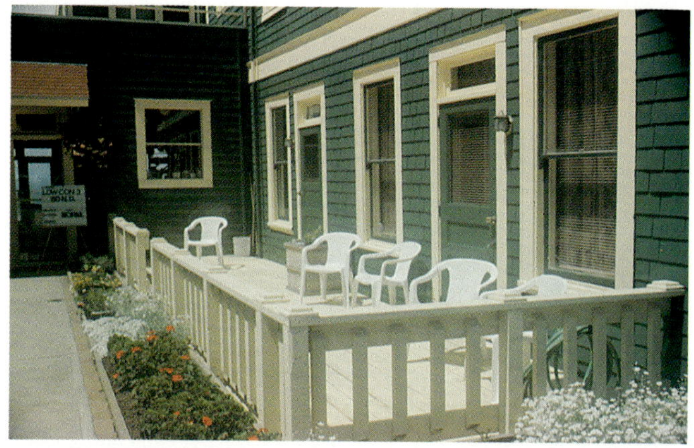

6D

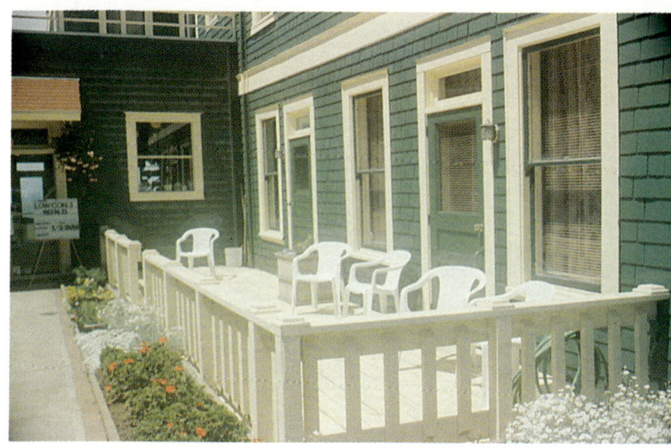

6G

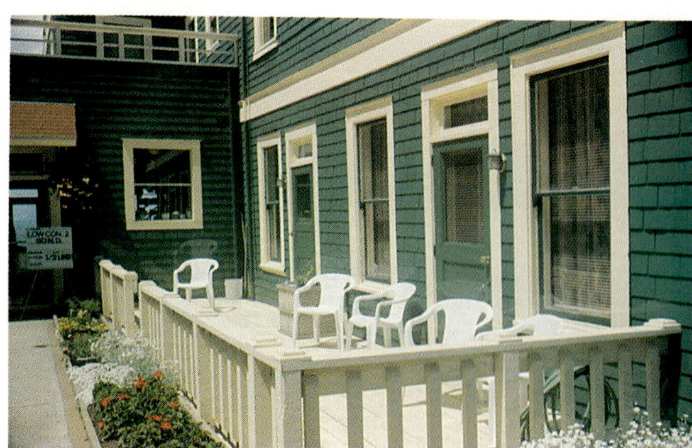

6J

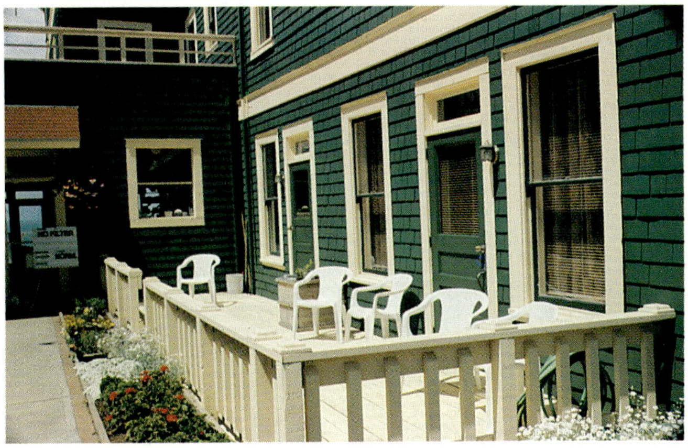

6B

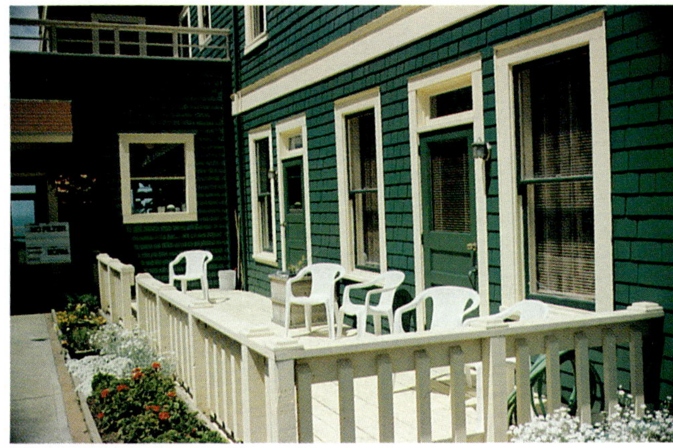

6C

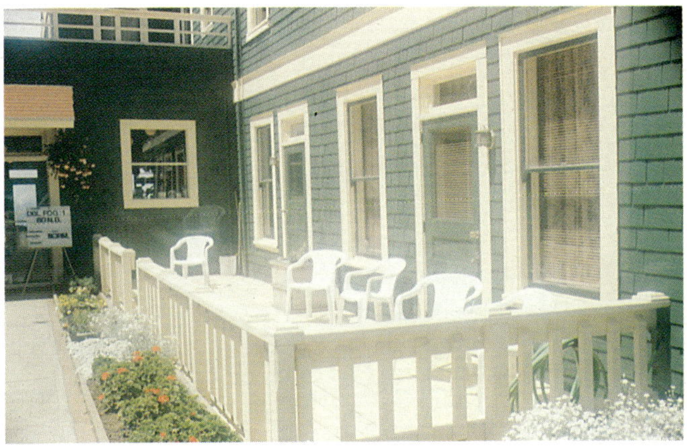

6E

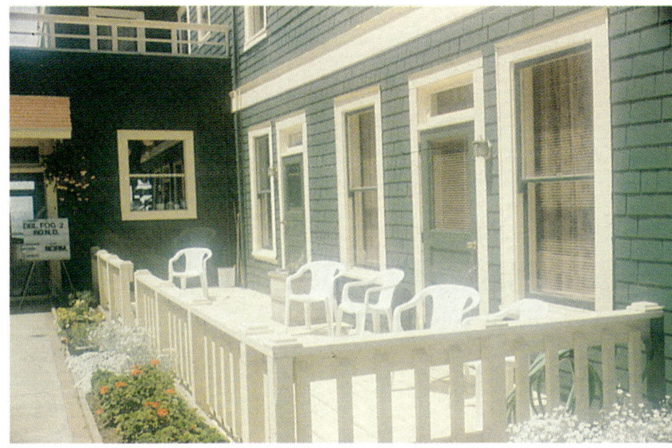

6F

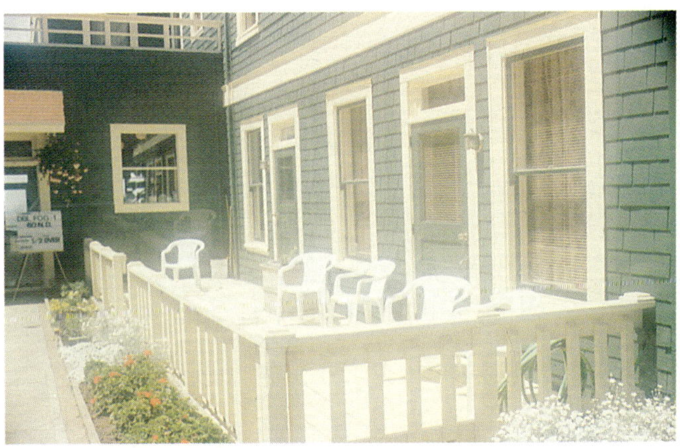

6H

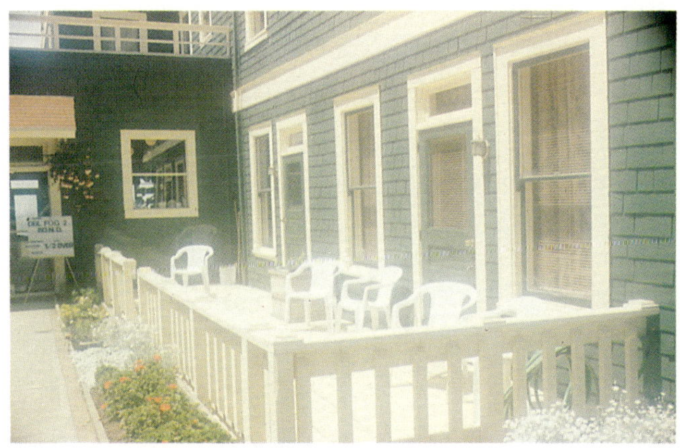

6I

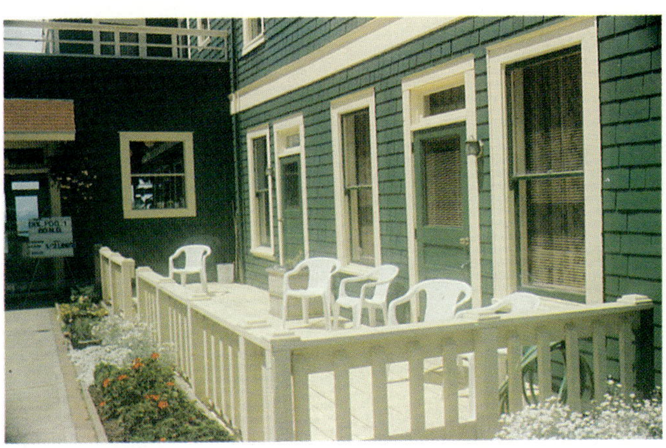

6K

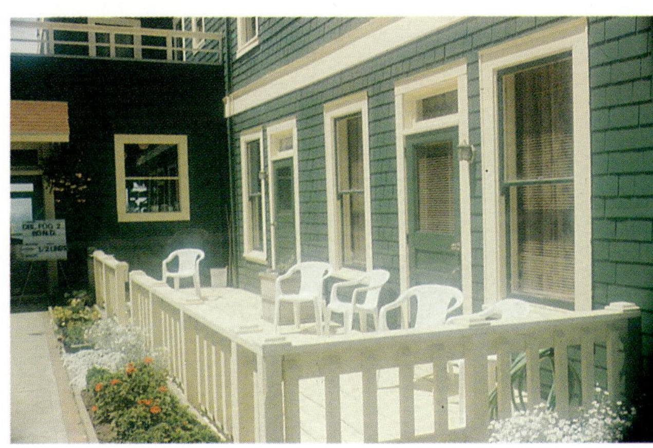

6L

PROTECT FILTERS FROM STRAY LIGHT (see page 21)

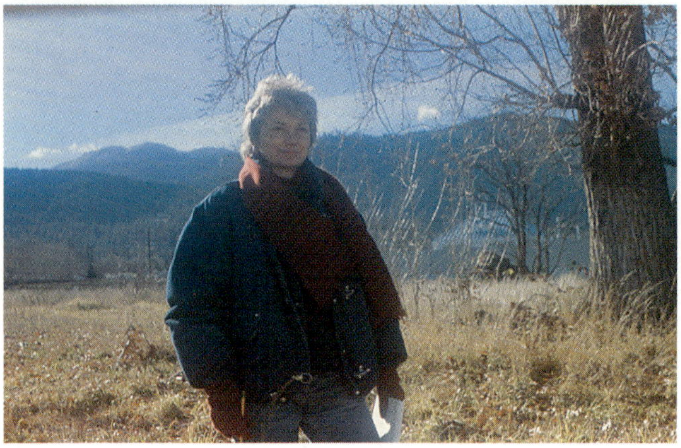

7A

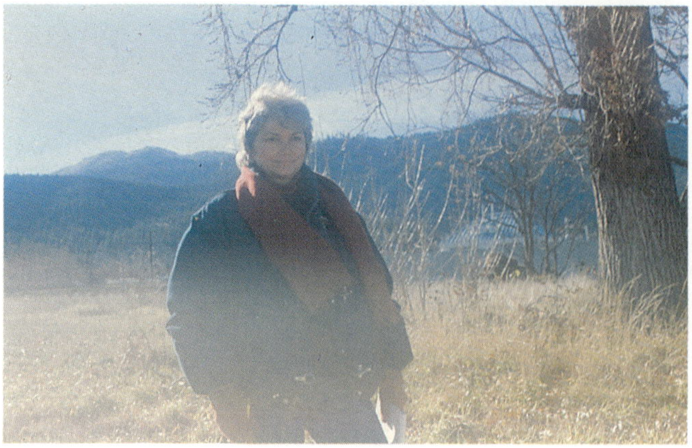

7B

Color Plate 7 A backlit scene made with a double fog #1, is shown in Plate **7A**. Compare this with Plate **7B**, in which the same scene was shot with the same filter, but this time the filter was not properly shielded from the sun. Although the sun did not hit the lens, the scene is spoiled because the sun hit the filter. Even on interior sets, the assistant cameraperson must always check that both the lens and the filter are properly shielded from light. A matte box is essential to protect the lens and the filters from extraneous light.

COMBINE FILTERS FOR GREATER VARIETY (see page 21)

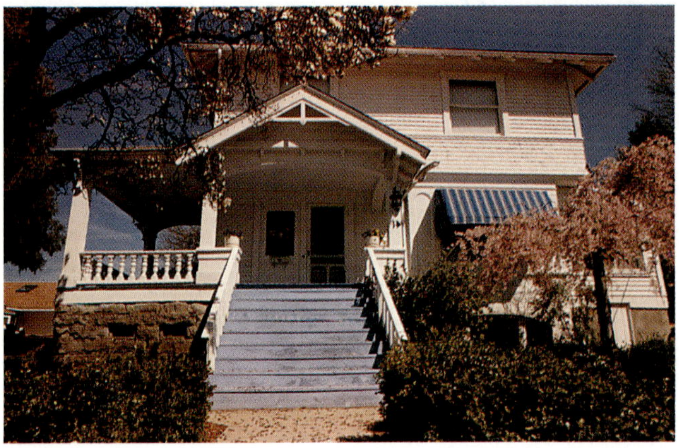

8A

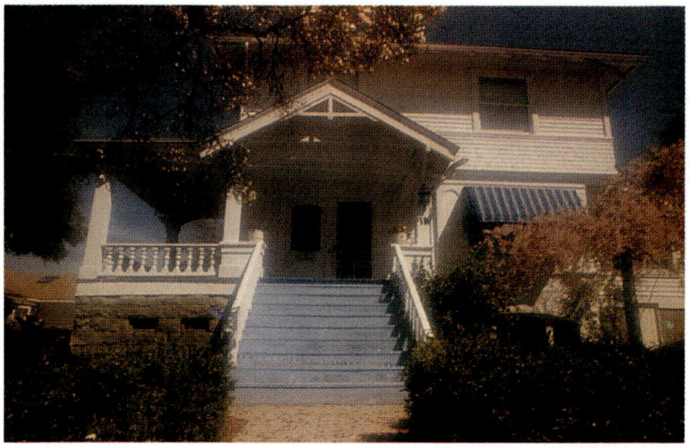

8B

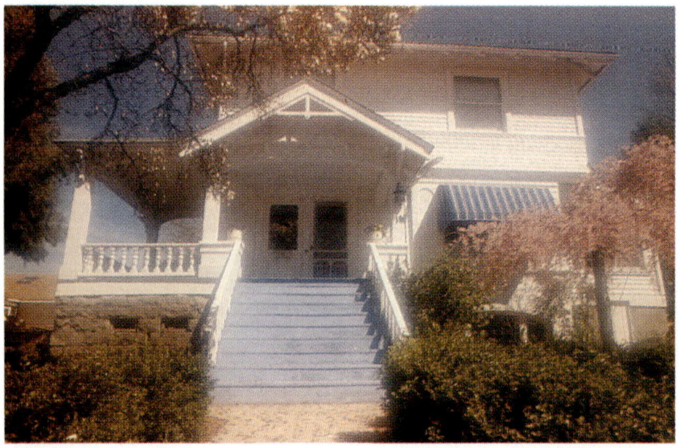

8C

Color Plate 8 Plate **8A** was shot with no filter and normal exposure. Note the saturated colors, and sharp detail. In Plate **8B**, made with regular fogs #3 and #1, and normal exposure, the colors are slightly muted, and the resolution is slightly softened. Plate **8C** was shot with regular fogs #3 and #1, and one-half stop overexposure. The colors are more muted and the image is softer. With a limited number of filters, combinations of two or more offer greater variety, and additional exposure control will accentuate the effect.

CREATING A FOG EFFECT (see page 24)

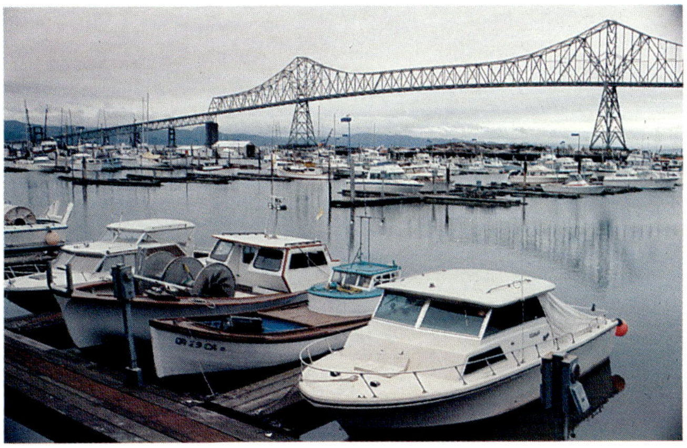

9A

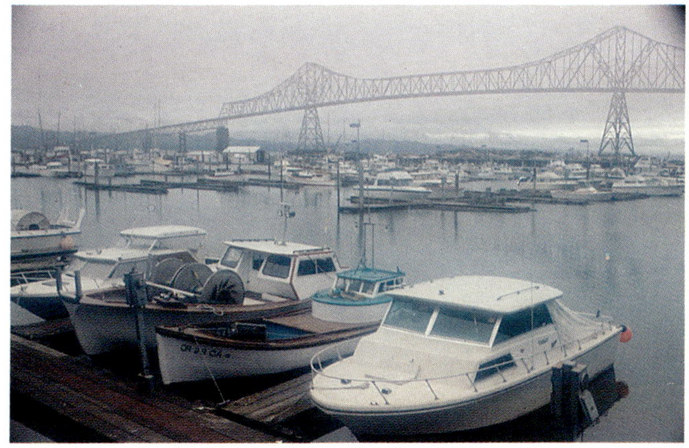

9B

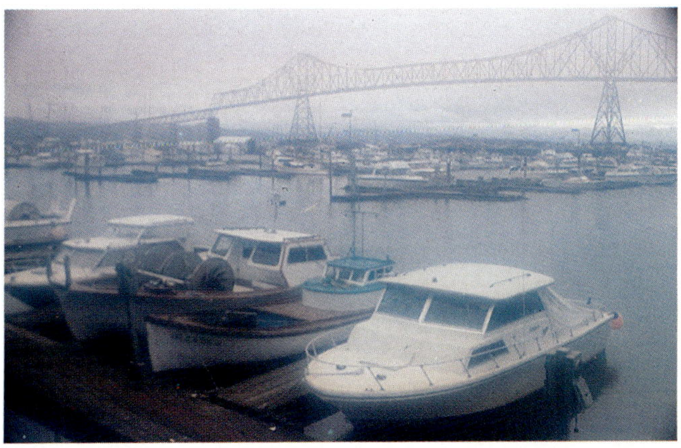

9C

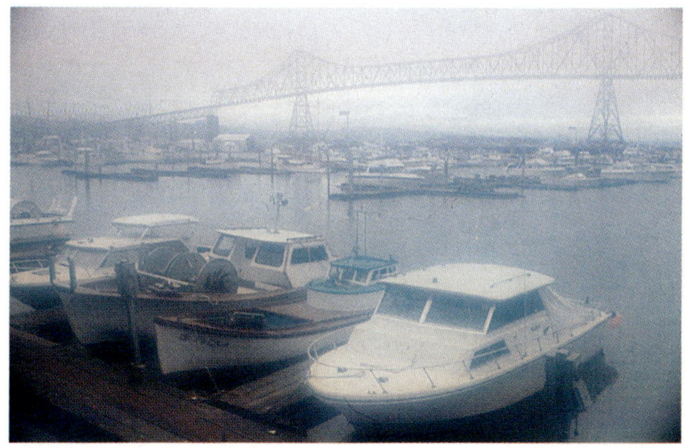

9D

Color Plate 9 Plate **9A** was shot with no filter and normal exposure. A scenic fog #3-to-clear filter was used for Plate **9B**, with normal exposure. The top of the scene was most affected. In Plate **9C**, a regular fog #2 was added, and the shot was overexposed one-half stop. The colors are more muted, the bridge is hazier, and the definition is softened. For Plate **9D**, the scenic fog #3-to-clear was combined with a regular fog #3, and the shot was over-exposed one stop. As the filter gets stronger, the effect increases. Overexposure also heightens the fog effect. Imagine how the scene would look if you were in a real fog and tried to simulate that look.

WHEN THE #85 FILTER IS NOT USED (see page 27)

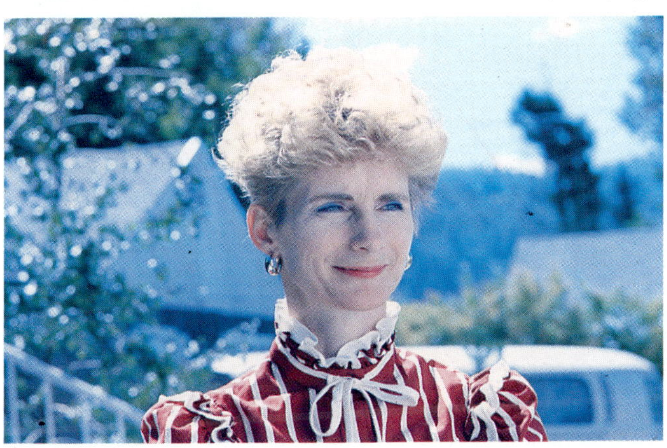

10A

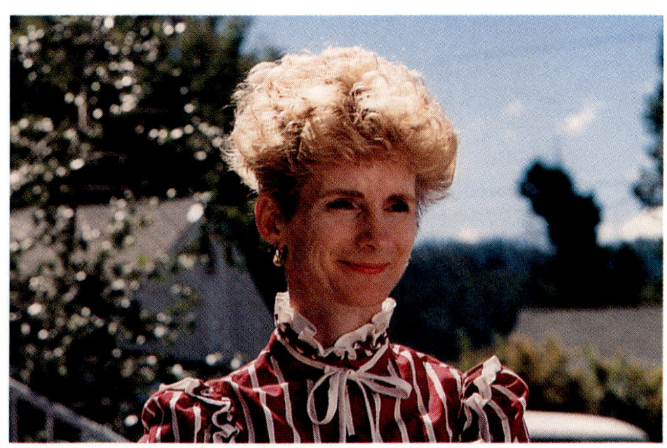

10B

Color Plate 10 The bluish scene shown in Plate **10A** resulted because the #85 filter was not used with a tungsten-balanced color negative (Eastman #5247) in daylight. If the light level is low and additional film speed is required, remove the #85 filter, and let the lab compensate by correcting the print. You gain two-thirds of a stop, and color is normal as shown in Plate **10B**.

THE CORAL FILTER (see page 29)

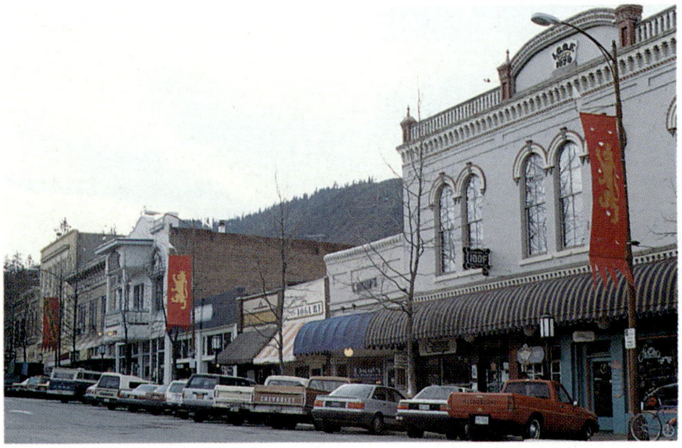

11A

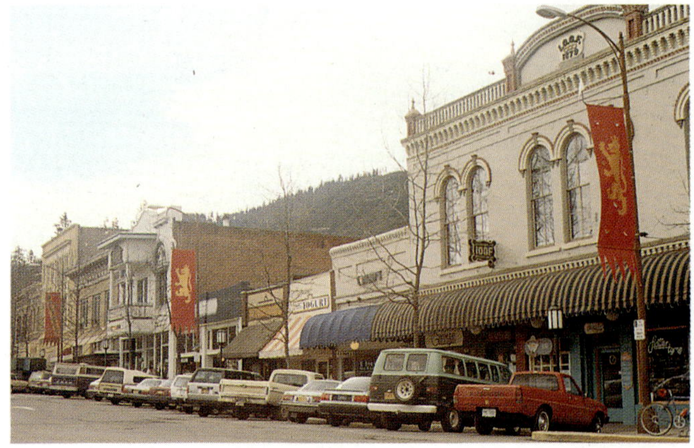

11B

Color Plate 11 A cold bluish tone is usually the result of filming in overcast. The color temperature may be as high as 10,000°K to 15,000°K, when normally it should be 5600°K. Plate **11A** was shot on an overcast day with only a #85 filter. Adding a coral #3 filter warms the scene for a more normal look, as shown in Plate **11B**.

CREATING A SEPIA TONE EFFECT (see page 30)

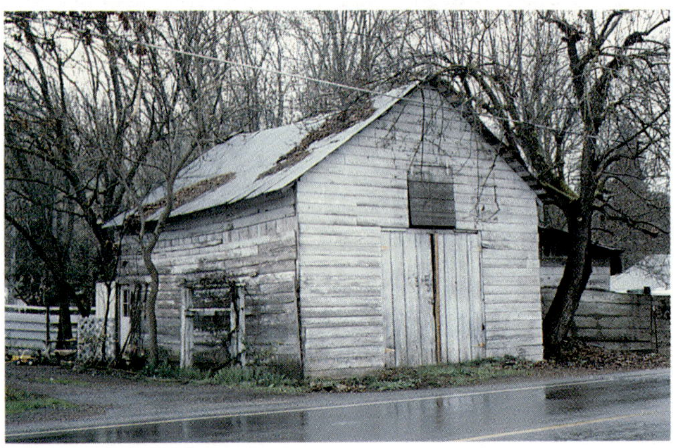

12A

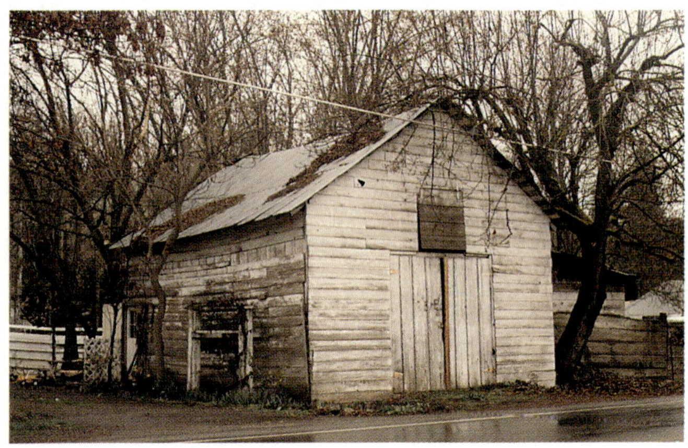

12B

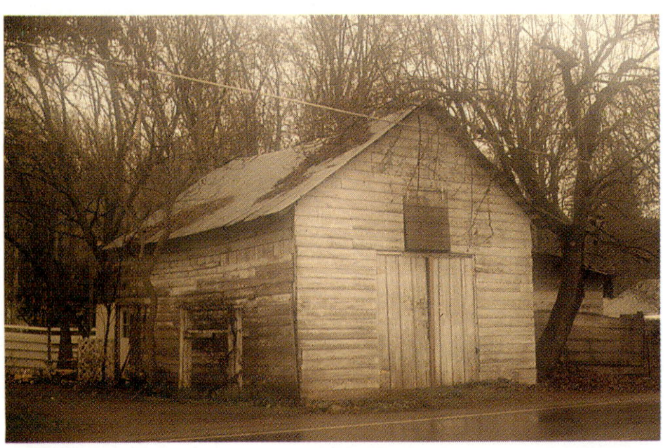

12C

Color Plate 12 Plate **12A** EK #5247 with #85, normal exposure; **12B** coral #5 (equal to a #85) added along with original #85. After compensating exposure for the added filter, the scene was underexposed one-half stop. **12C** #85 plus coral #5 plus regular fog #3 plus a "chocolate" filter (homemade from Rosco #682 Cinegel) a filter factor of 3 was determined for the chocolate filter. After all the filter factors were calculated, shot was made one-half stop underexposed.

GRADUATED FILTERS (see page 31)

Color Plate 13 These four filters are classified as graduated filters. The two on the left, H&H #85B-to-clear (above) and #B4-to-clear (below) are hard-edged graduated filters. The one at the upper right, a .60ND-to-clear, is a soft-edged graduated filter. At the bottom right is a one-stop attenuator (NBRA), which gradually changes from clear at the bottom to a .60ND at the top. Both of the NDs were made by Pancro-Mirror, Inc.

THE STRIP FILTER (see page 31)

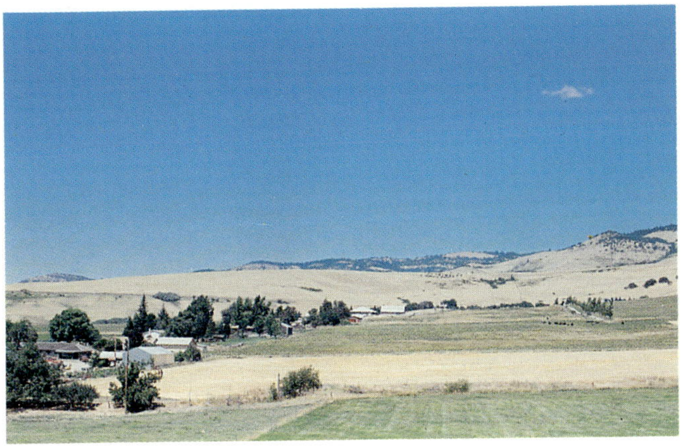

14A

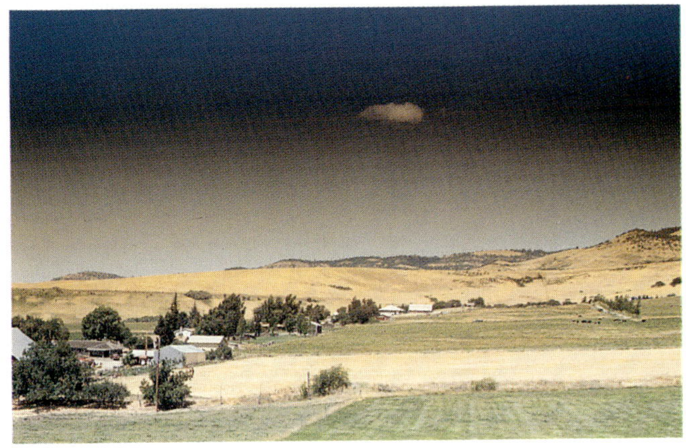

14B

Color Plate 14 Plate **14A** shows an unfiltered landscape. The same scene made more dramatic in Plate **14B** by using a 60ND-to-clear on the upper third of the frame and an amber strip filter just below it. No exposure compensation is required when the main part of the scene is shot through the clear portion of the filter.

GRADUATED FILTERS: VIEW THROUGH THE LENS STOPPED DOWN (see page 31)

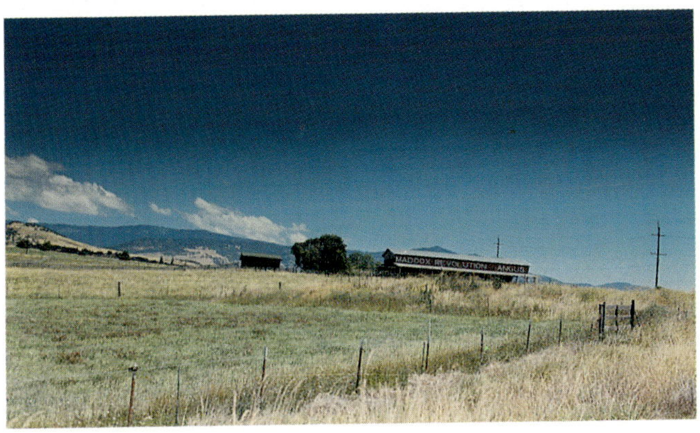

15A

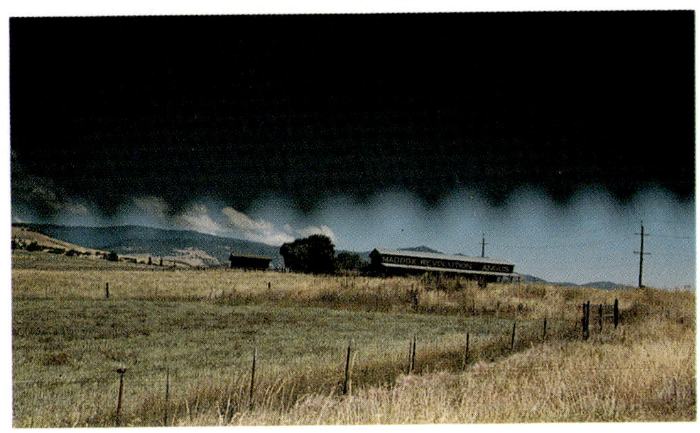

15B

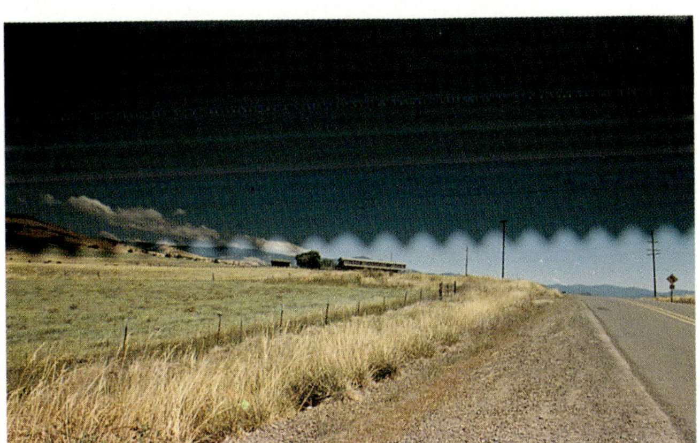

15C

Color Plate 15 A soft-blend homemade filter using pinking shears to cut the edge of the colored gel, was used for Plate **15A**. The filter might be more effective if it were in a matte box with a rotating filter holder to tilt it to conform with the horizon line. The same filter was used in Plate **15B**, but the lens was stopped down more. The saw-toothed edge is sharper, spoiling the shot. A finer saw-toothed edge or a manufactured filter might have prevented this effect. In Plate **15C**, the same filter was used, but the lens was changed to a wide angle and stopped down. This shot too is unusable. Always view the effect of any graduated or half filter with the lens stopped down to the exposure to be used, regardless of the focal length of the lens.

SPECIAL EFFECTS FILTERS (see page 32)

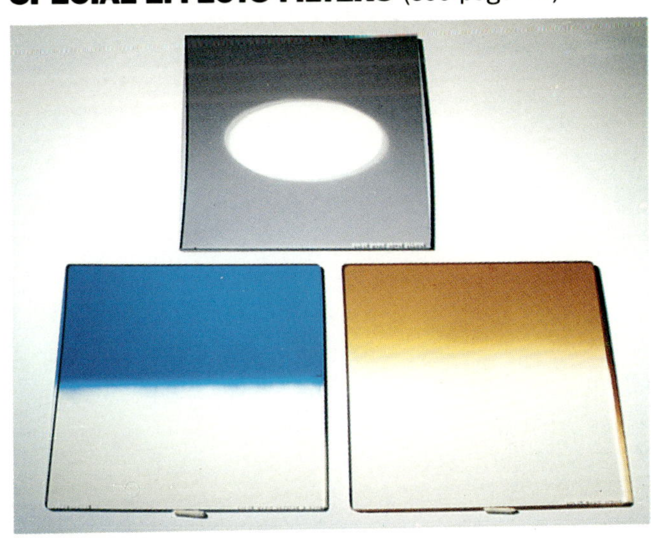

Color Plate 16 These three filters are 6.6" square, they can be placed with the dividing line vertical or horizontal, and may be rotated in the proper matte box. Their size will cover the angle of most lenses. These filters are made of cast resin. As such they are a solid piece, not two pieces of glass taped on the edges, and are manufactured by Color Magic. *Clockwise from top:* an ellipse matte for vignetting, an orange graduated filter with a soft blend line, and a blue graduated filter with a hard blend line.

FILTERS MAKE A SCENE MORE DRAMATIC (see page 32)

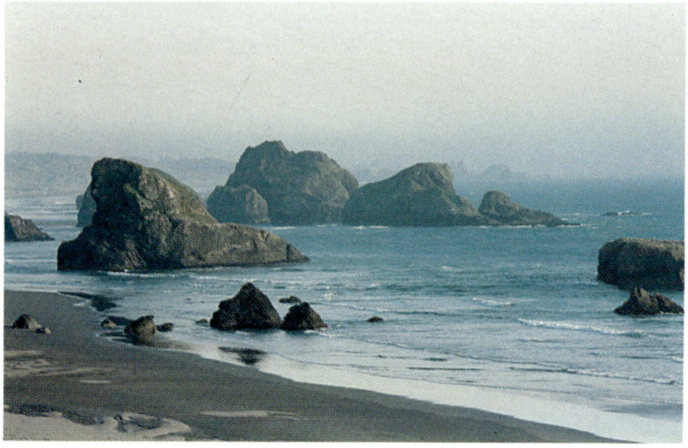

17A

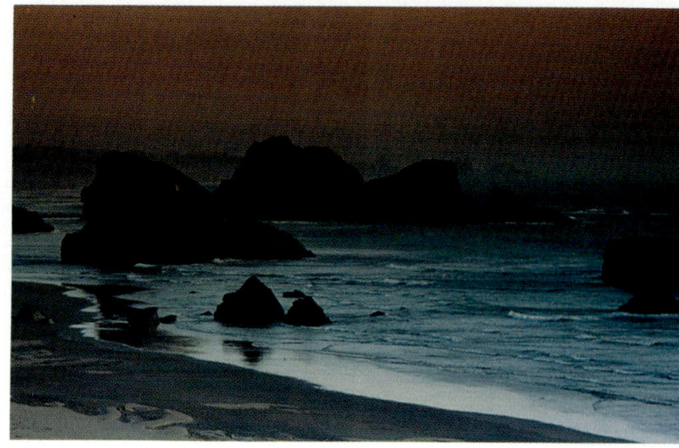

17B

Color Plate 17 A normal seaside scene, (Plate **17A**), is dramatically changed through the combined addition of a #85B-to-clear filter and a .90ND-to-clear filter (Plate **17B**). No exposure compensation was made.

USING DIFFUSION (see page 41)

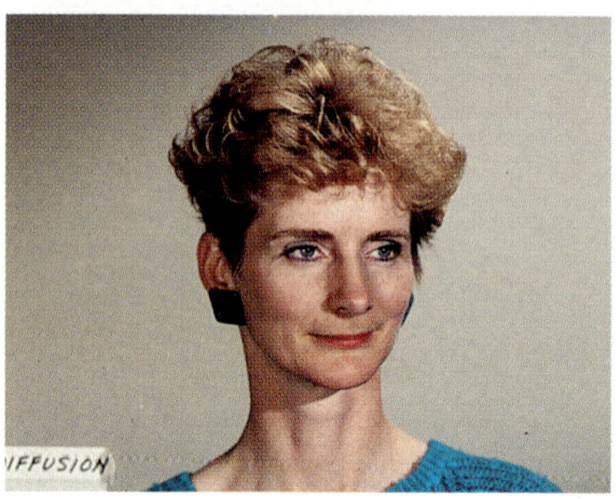

18A

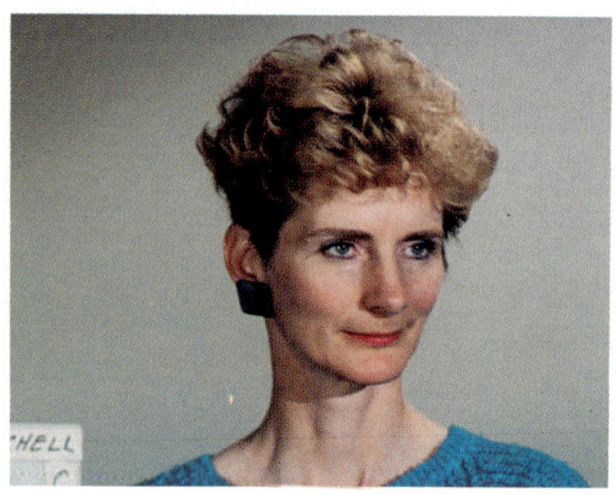

18B

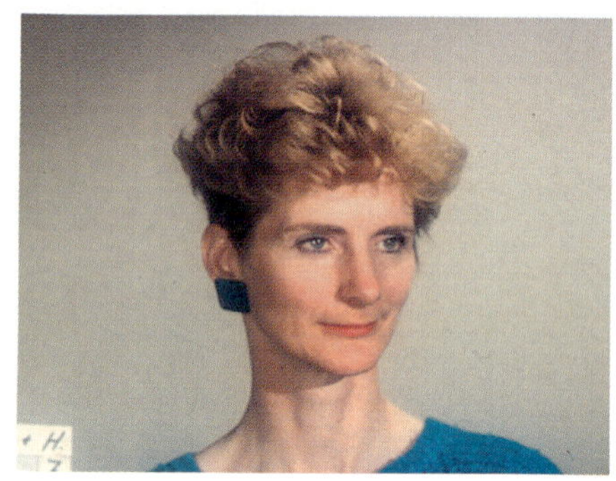

18C

Color Plate 18 Plate **18A** was shot without diffusion. The diffuser used for Plate **18B** Mitchell C (Pancro-Mirror) softened the lines but did not affect the contrast. In Plate **18C**, the H&H D-3 filter softened the lines and also reduced the contrast. H&H also makes a black dot diffusion filter, called *texture screen,* that softens the image and retains contrast. If you use Mitchell diffusers and want to reduce contrast, you may add a low-contrast or fog filter. Reduced contrast is sometimes desirable to help hide facial blemishes. You may occasionally want to retain the contrast but reduce the resolution of the lens.

DIFFUSION EFFECT (see page 41)

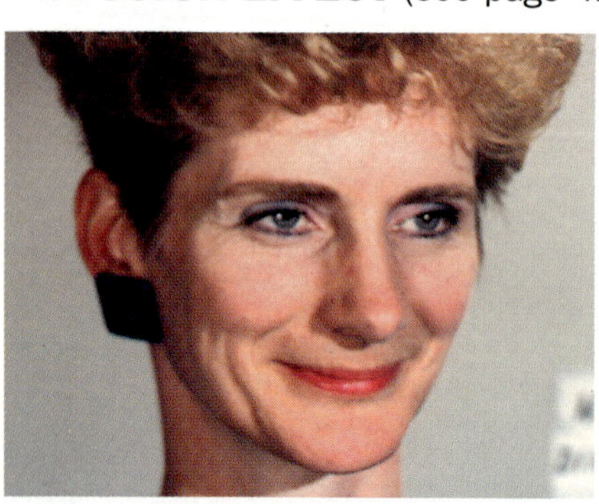

19A

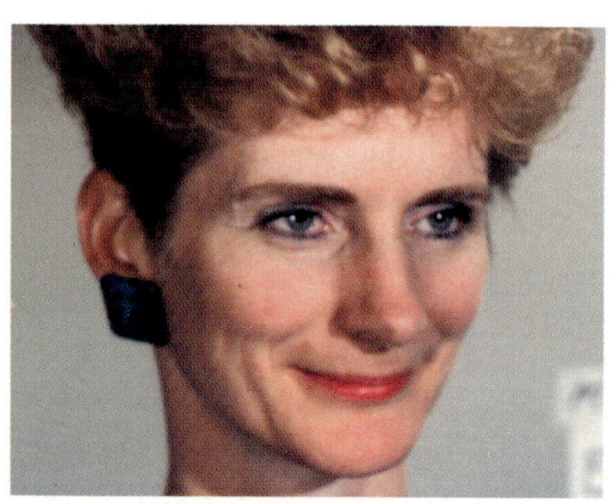

19B

Color Plate 19 Extreme close ups call for heavier diffusion. Plate **19A** used an H&H D-2 diffuser. Plate **19B** used a Mitchell "E." The choice of which to use will depend on the dramatics of the screenplay.

HOMEMADE FILTER EFFECTS (see page 51)

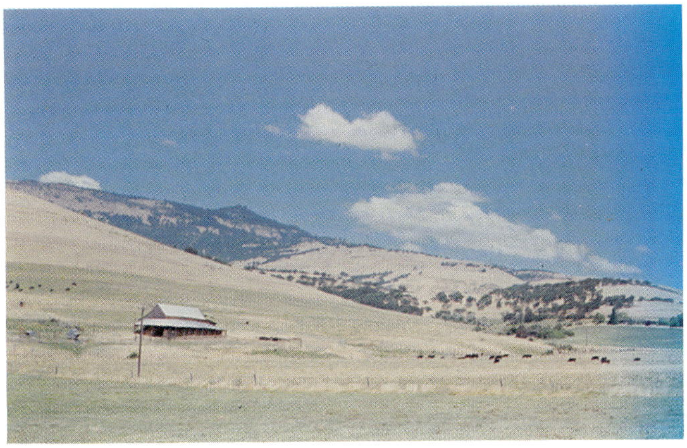

20A

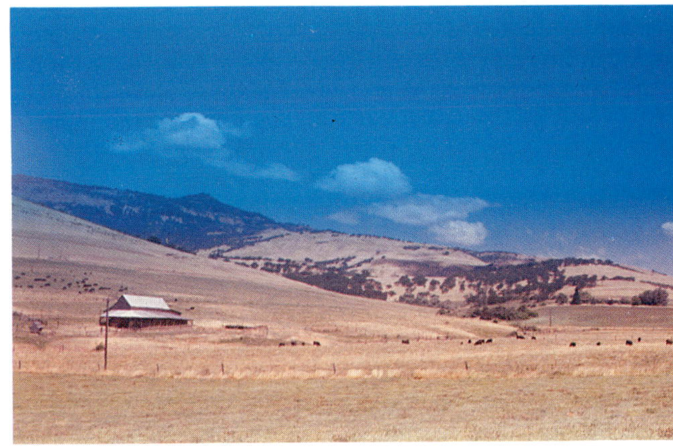

20B

Color Plate 20 No filter was used for Plate **20A**. In Plate **20B** we used a homemade filter with blue on top and rose below, creating a unique landscape. An exposure meter was used with the rose-colored gel over the meter's cell to determine the filter factor.

DAY-FOR-NIGHT EFFECT (see page 51–52)

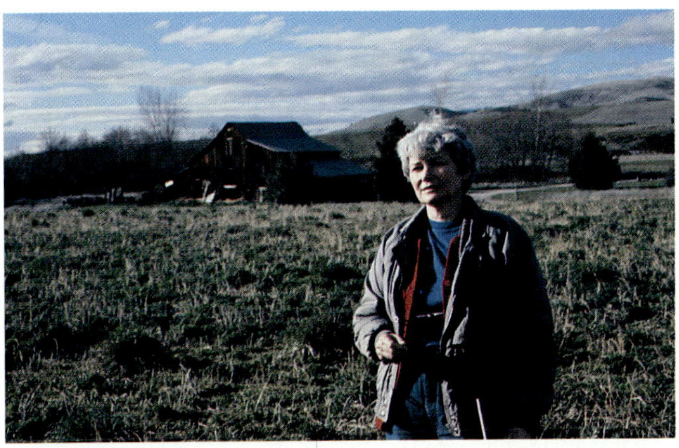

21A

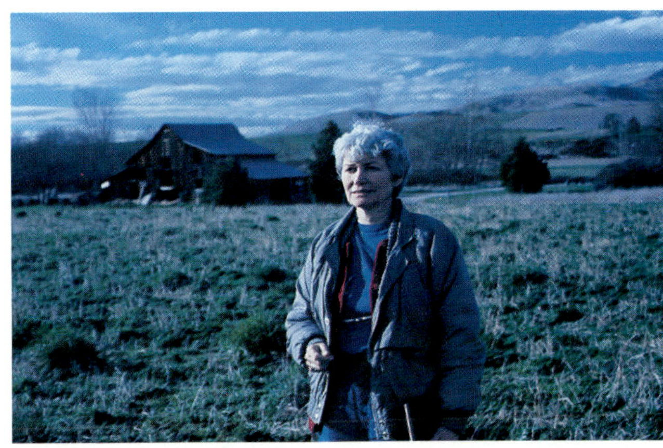

21B

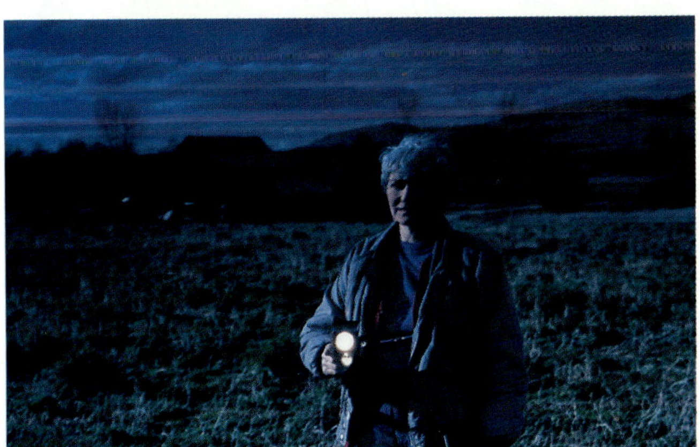

21C

Color Plate 21 A normal day shot, Plate **21A**, was made with tungsten film, a #85 filter, and normal exposure. For Plate **21B**, the filter was removed, a polarizer was added for darker sky, and the shot was exposed normally. The day-for-night effect is shown in Plate **21C**. The #85 was removed, a polarizer and .90ND-to-clear filter were used to further darken the sky. The shot was underexposed, one and a half stops and filmed in cross light for shadow areas. A flashlight (100 watt quartz bulb in reflector) was pointed at the lens to heighten the night feeling. A good day-for-night effect requires that sunlight become moonlight.

LAB TEST SLATE (see page 67)

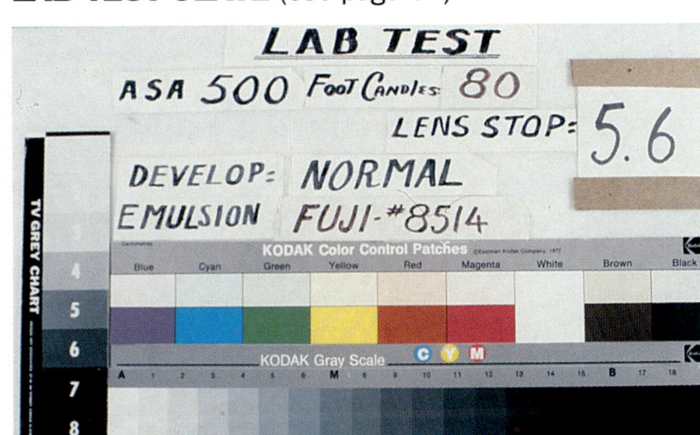

Color Plate 22 The information card for the definitive lab test. The lens stop numbers can be slid in and out for easy changes. Information stating the amount of over- or underexposure can be added.

LAB TEST ONE-LIGHT PRINT (see pages 11, 67–68)

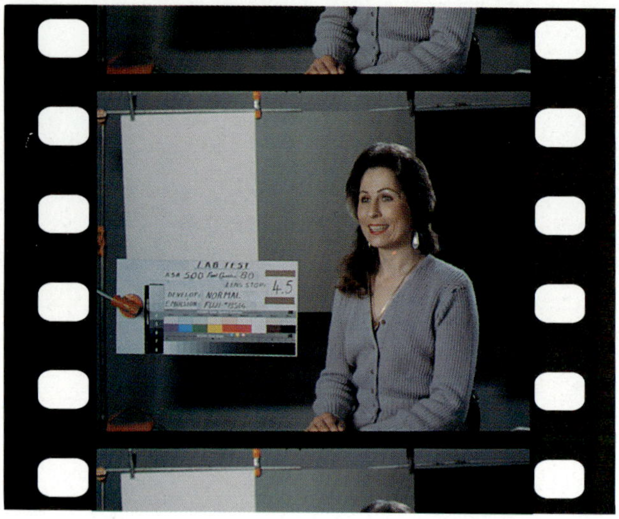

F/4.5 1/2 stop OVER

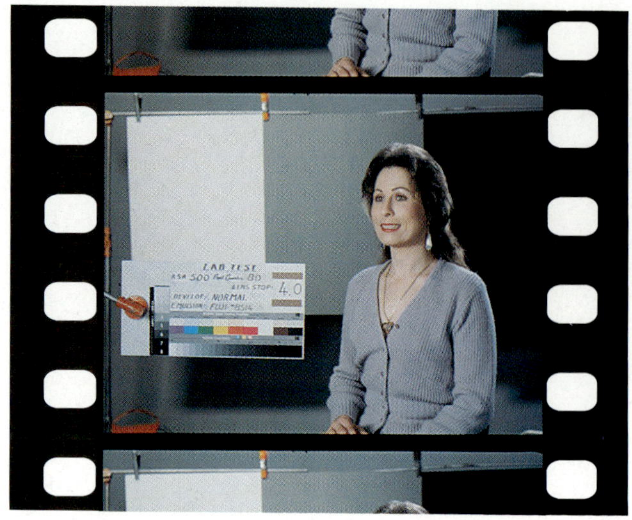

F/4 1 stop OVER

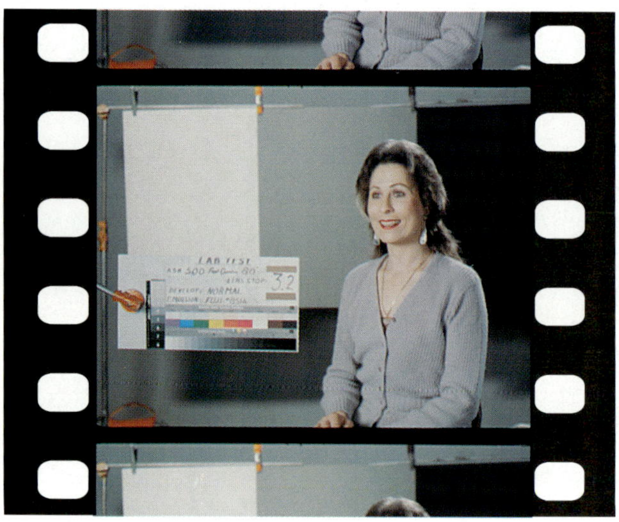

F/3.2 1-1/2 stops OVER

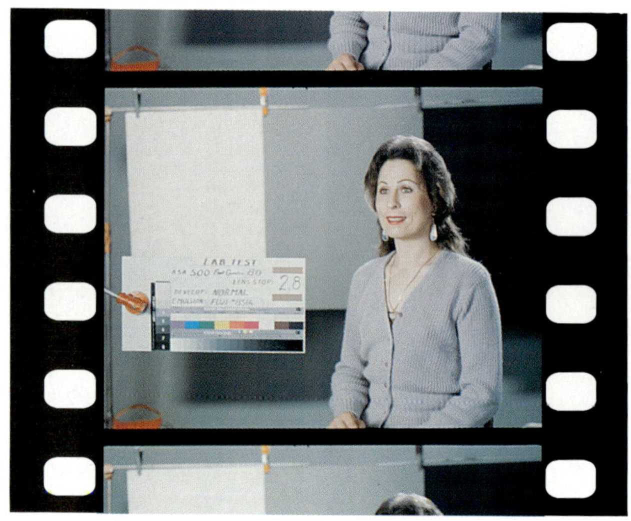

F/2.8 2 stops OVER

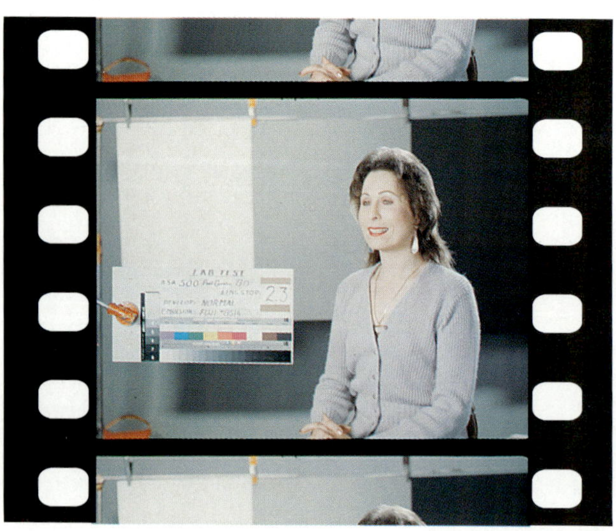

F/2.3 2-1/2 stops OVER

Color Plate 23 Lab test, one-light prints. This lab test was made using Fuji #8514 and processed at Technicolor Lab in California. The exposure index for this emulsion is 500, and the lab was instructed to process it normally and to make a one-light print based on the normal exposure of F/5.6. The lab is asked to use their normal printer lights to see if their processing affects the exposure index. In which case, the cinematographer would alter the EI for that particular lab. An exposure range of five stops was used for this test. Note that all the pertinent information about each exposure is filmed within the frame. The printer light numbers were 40 points of magenta, 40 points of cyan, and 22 points of yellow. It may seem like a large difference between the yellow and the magenta and cyan, but it was necessary to render a good flesh tone with these particular negative print emulsions.

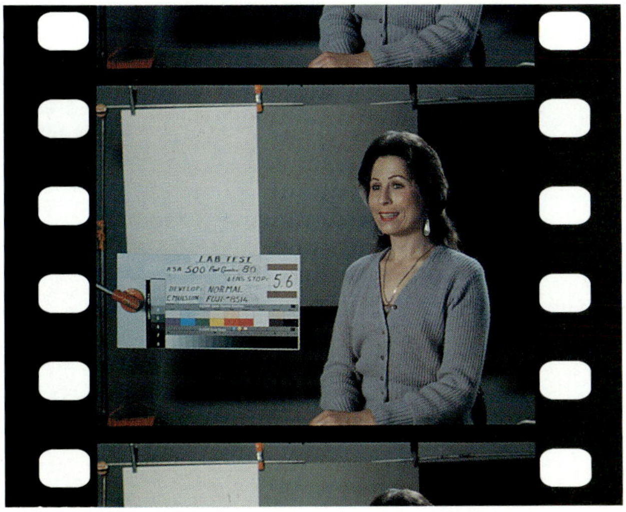

F/5.6 NORMAL EXPOSURE

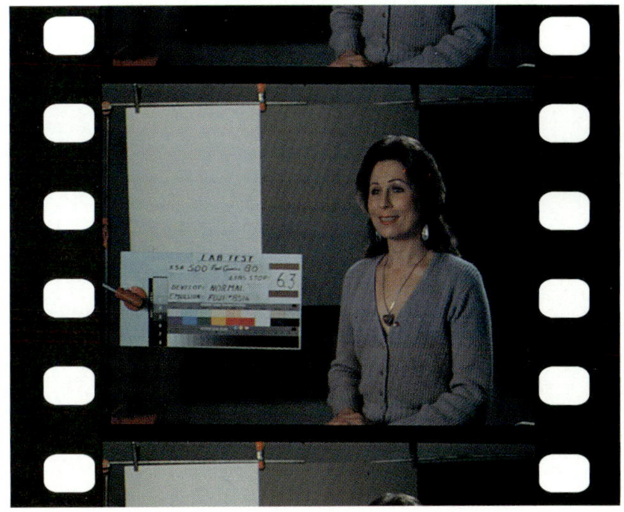

F/6.3 1/2 stop UNDER

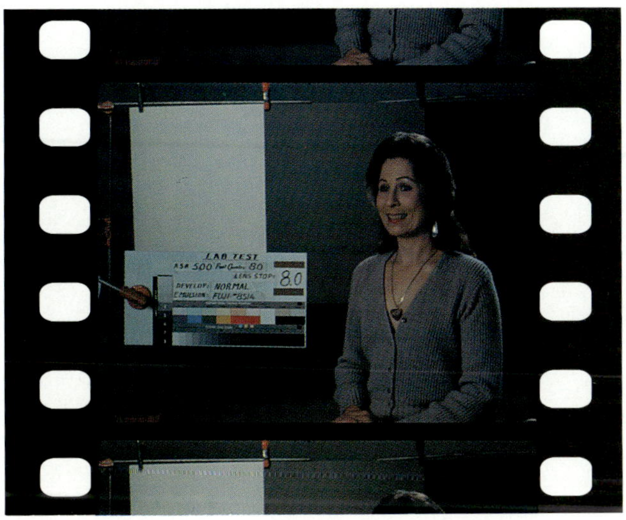

F/8 1 stop UNDER

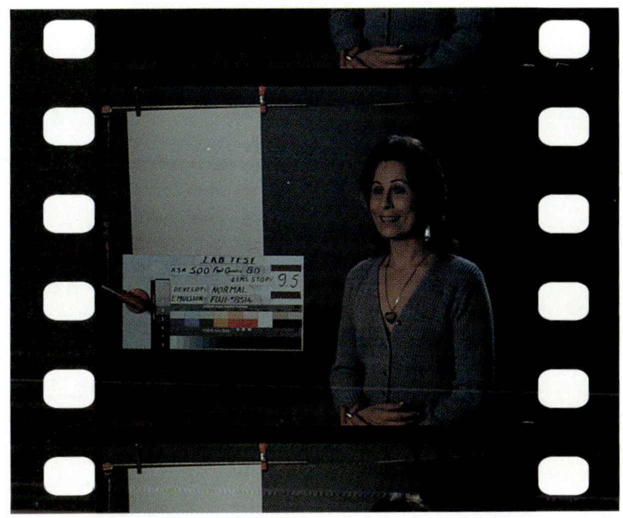

F/9.5 1-1/2 stops UNDER

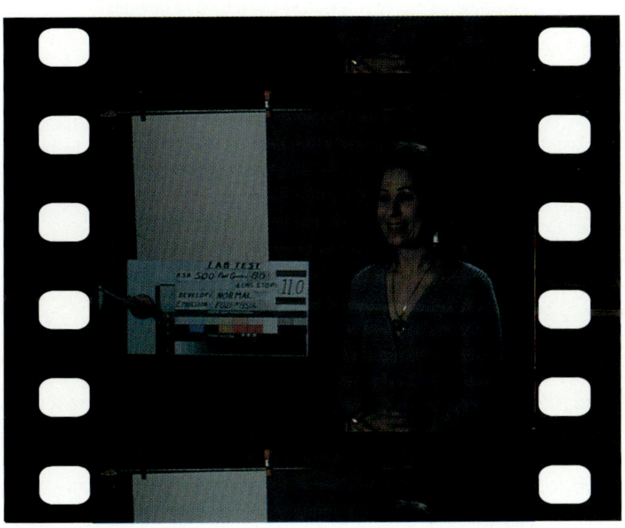

F/11 2 stops UNDER

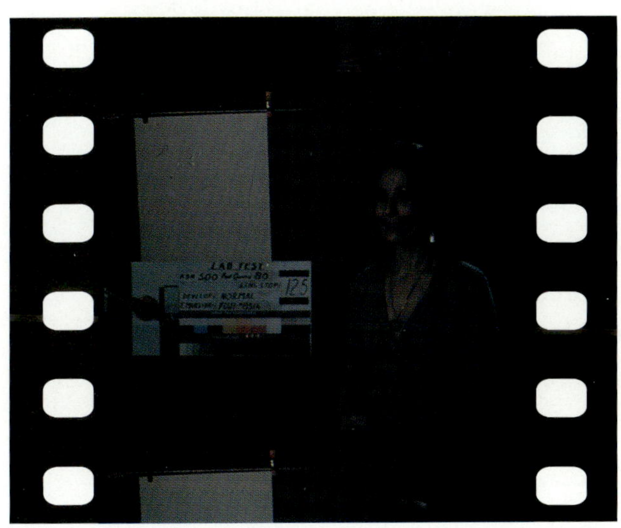

F/12.5 2-1/2 stops UNDER

LAB TEST TIMED PRINT (see pages 11 and 69)

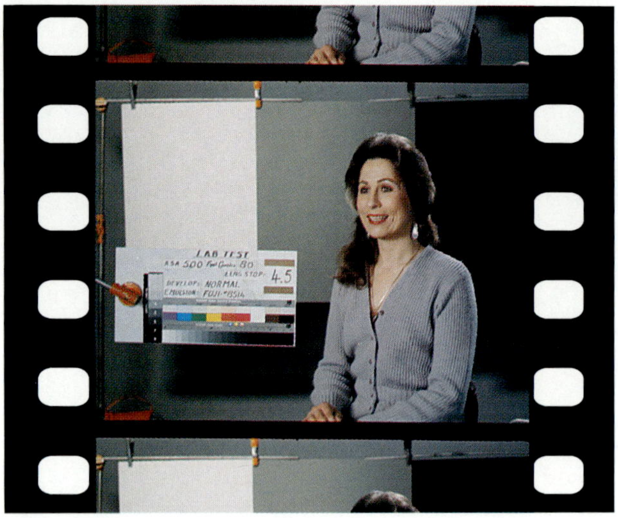

F/4.5 1/2 stop OVER, M42, C42, Y24

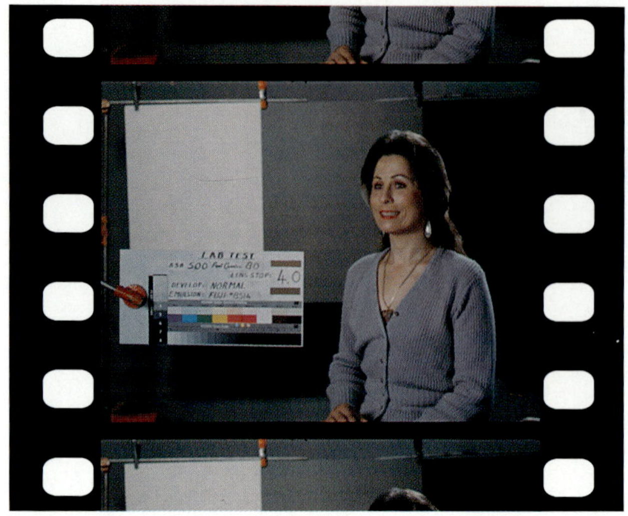

F/4 1 stop OVER, M46, C45, Y29

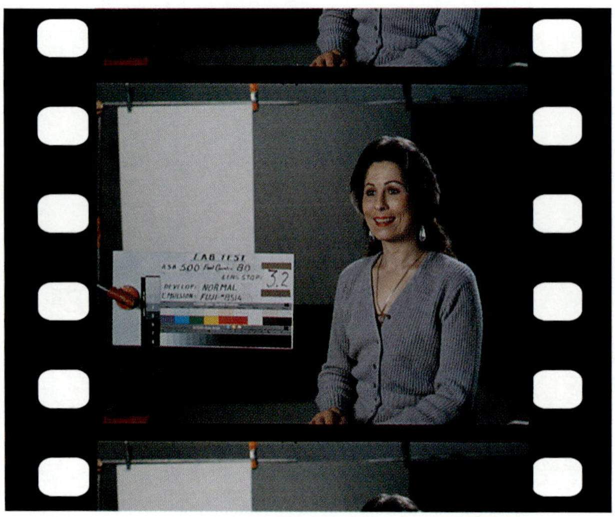

F/3.2 1-1/2 stops OVER, M48, C47, Y32

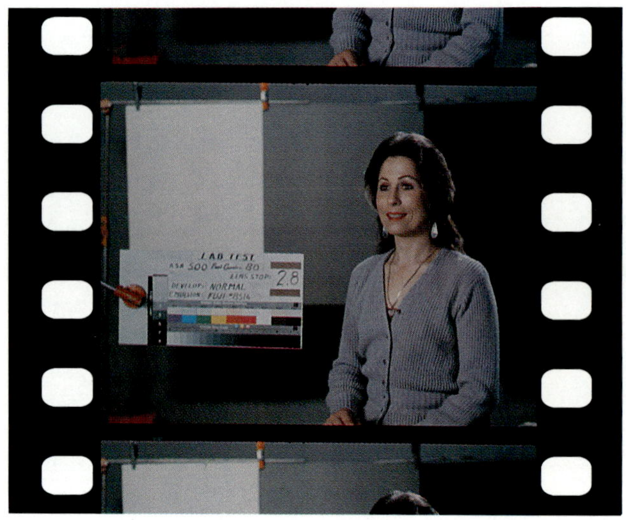

F/2.8 2 stops OVER, M50, C48, Y33

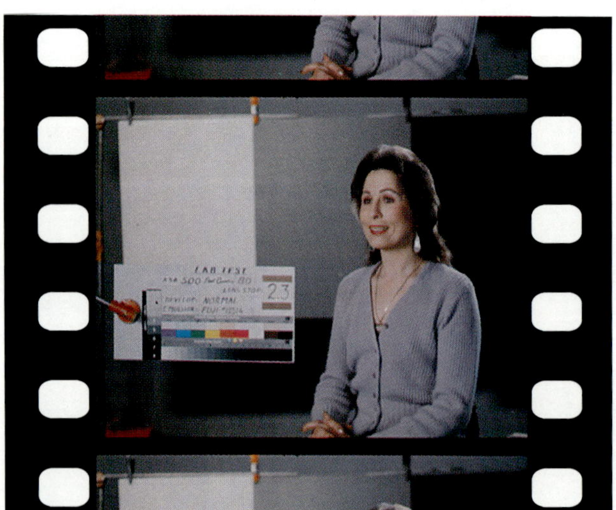

F/2.3 2-1/2 stops OVER, M50, C48, Y33

Color Plate 24 Lab test, timed prints. Using the same color negative that was used for the one-light print, the lab was instructed to make a "timed print" (the best possible print from each of the different exposures which covered a range of five stops). This test shows that a good normal print can be made anywhere from two stops overexposed to one and a half stops underexposed. Note: When overexposed by two stops and by two and a half stops, F/2.8 and F/2.3, the printer lights were the same. The reason for this is that 50 is the top of the printer scale for printing a dense negative, and although the yellow and cyan could be printed darker, it would throw the color balance off, so the timer left the lights the same for both exposures. The exposure index could be changed to 1000 if needed. By examining the printer lights, you can see that at a stop of F/8 (underexposed one stop) the lights were 33, 34, 15. These are relatively in the middle of the printer scale, but you must be careful that the yellow printer light (15), representing the blue component of the color emulsion, doesn't get so low that there will be a loss of the "black" density. This is evident in the exposures made at F/11 and F/12.5, where the yellow printer lights are only 8 and 6, respectively, and the blacks are no longer solid black. Processing by another lab could show different characteristics regarding the exposure index, contrast, and grain for a normal density negative.

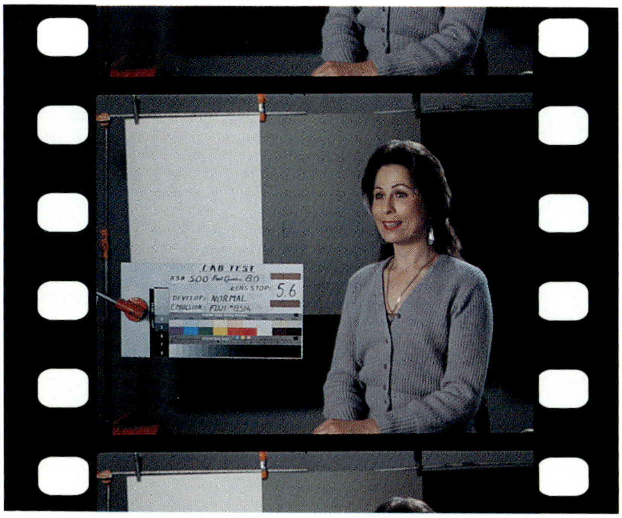

F/5.6 NORMAL EXPOSURE, M40, C40, Y22

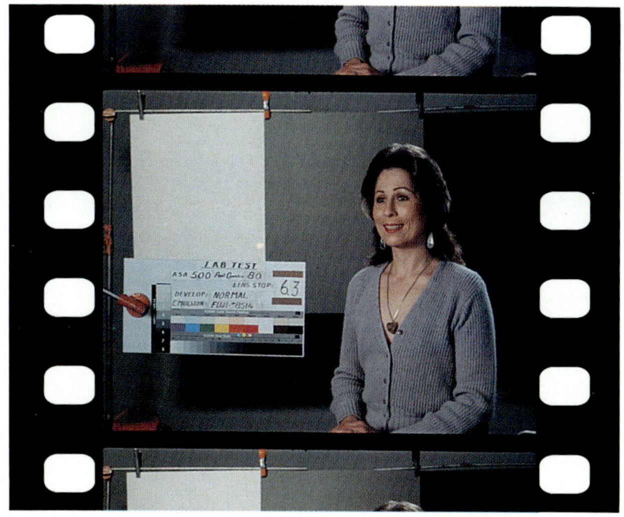

F/6.3 1/2 stop UNDER, M36, C37, Y18

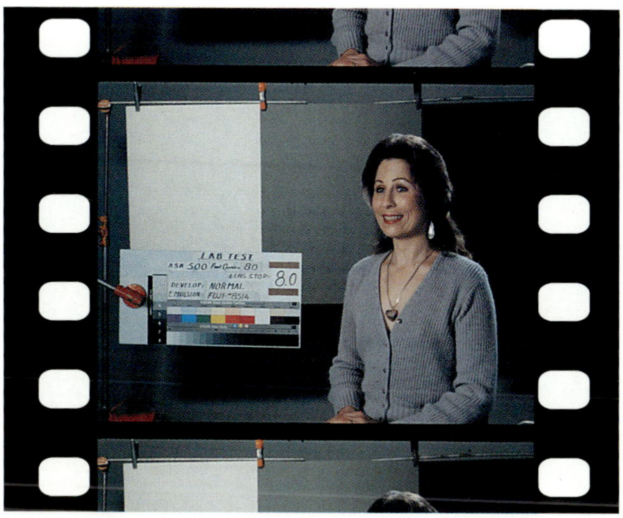

F/8 1 stop UNDER, M33, C34, Y15

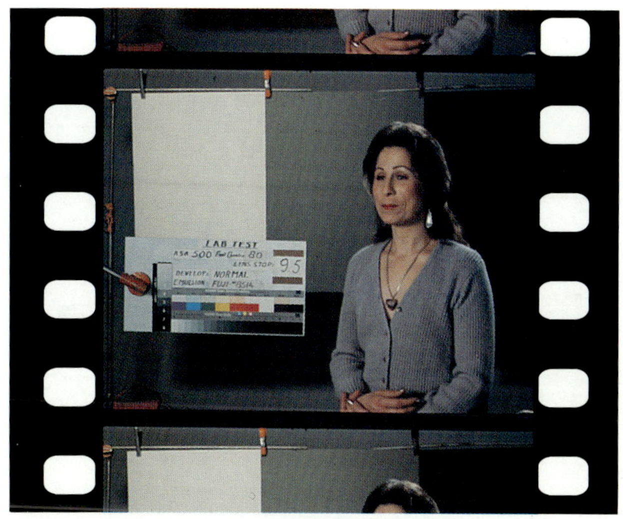

F/9.5 1-1/2 stops UNDER, M30, C31, Y11

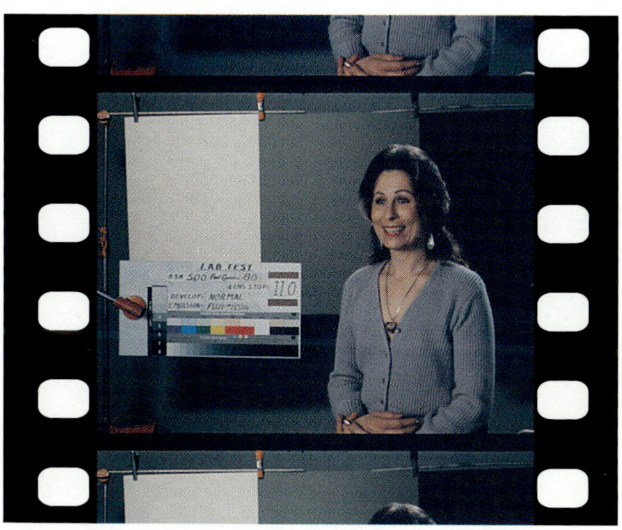

F/11 2 stops UNDER, M27, C28, Y8

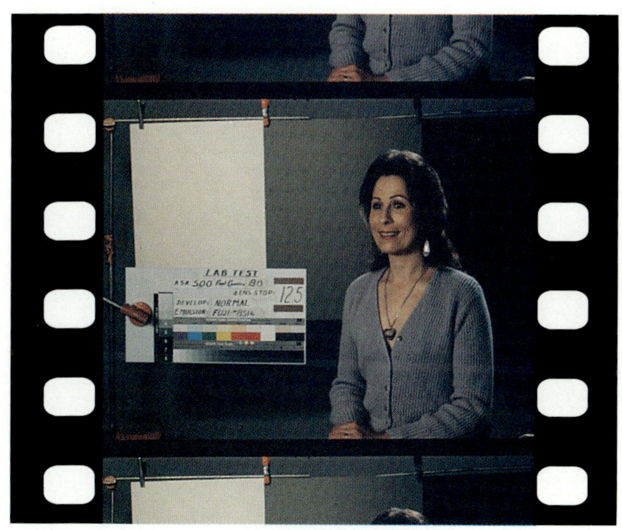

F/12.5 2-1/2 stops UNDER, M24, C25, Y6

NET FILTERS (see page 42)

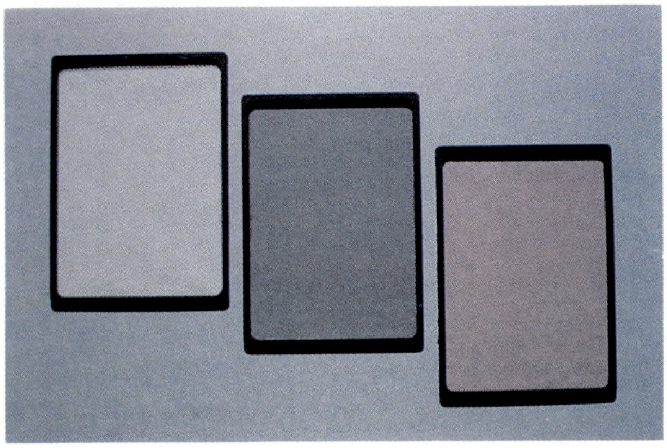

Color Plate 25 Three samples of net filters are shown. The white net diffuses and reduces contrast; the black net only diffuses; and the red net diffuses and adds a touch of tone to the image. These nets are mounted in frames made by Panavision. Frames are also used to hold gel materials. Tiffen makes net diffusers that are permanently mounted in plastic.

STAR FILTER COMBINED WITH COLOR FILTER (see page 44)

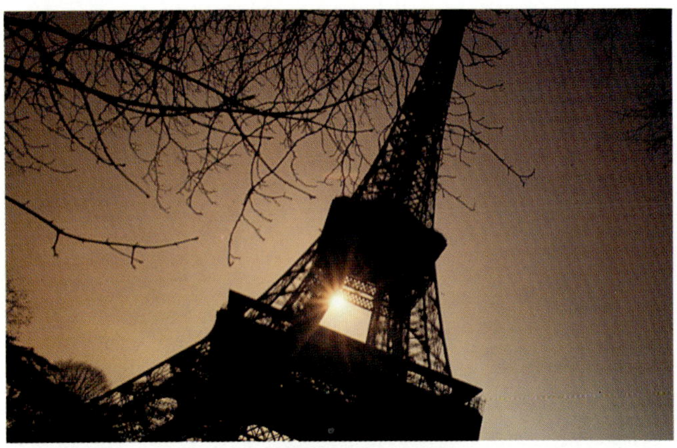

26A

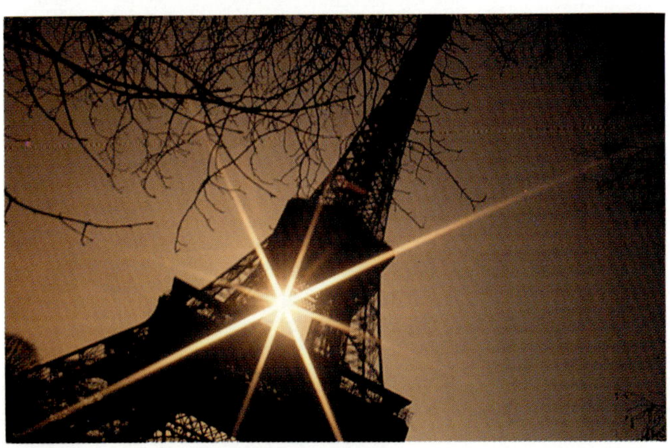

26B

Color Plate 26 Plate **26A** was shot with Cokin Filter #198 sunset filter. The same filter was used in Plate **26B** with a Cokin filter #056, eight-point star filter. (Courtesy of Cromofilter S.A.)

MAKING A GRADUATED FILTER

(see pages 50–51)

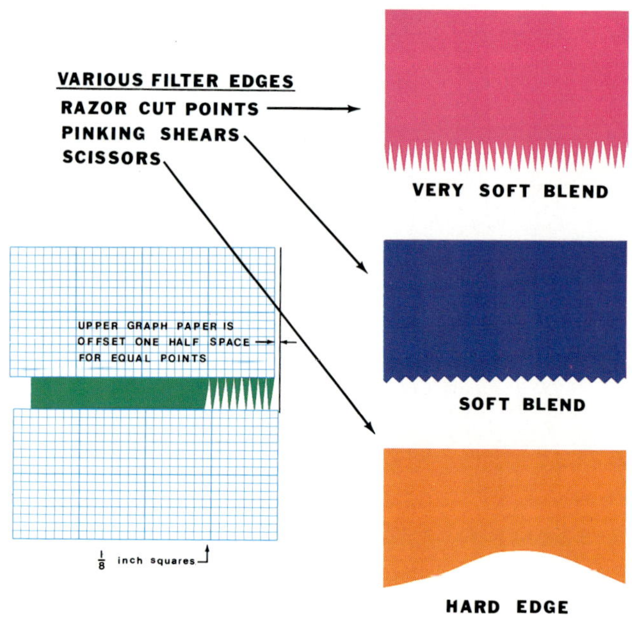

Color Plate 27 Illustrated are the different means of making edge blends for homemade filters and the tools used.
Top: A filter cut with a razor blade with long points about one-eighth-inch apart creates a very soft blend.
Middle: A filter cut with pinking shears, which results in a moderately soft blend, requires careful checking when the lens is stopped down.
Bottom: A filter with a hard edge made with scissors is used for special applications only.

are filters for use in sunlight, day interiors, night exteriors, studio fill, candlelight, sepia, various period effects, and golden exterior. Imagine how effective the process would be if you had to film in a pine forest and make very expansive shots. All you'd have to do to give the background more detail and have it record as natural green instead of black, would be to dial in some green light through the Lightflex! The amount of light that's added to the shadow areas does not affect areas of normal exposure such as faces. (See Figure 11.)

When would you use the Lightflex? As always, the screenplay is the first consideration. After contemplating its content and visualizing the look that will complement and support the characters, you can then plan to use techniques that will alter the normal screen image in ways that are beneficial to the telling of the story. When you've become familiar with the many uses of the Lightflex, you'll be in a position to suggest its use and to employ it as an adjunct to your lighting. Because it affects shadow areas and not highlights, the Lightflex can bring details that would otherwise be lost into the printing range. It can be used either to reduce the scene contrast or to overlay a subtle color on the scene. You're able to judge the results right then and there on the ground glass, something you cannot do when using negative flashing techniques. A word of caution: if improperly used, the Lightflex can spoil the look of the image just as too heavy a filter can. As with any camera accessory or lab technique, make complete tests of the Lightflex before putting it to work.

The complete Lightflex system consists of the following: the main matte box unit, two internal curved reflectors—one for focal lengths of 14mm to 50mm and the other for 75mm to 100mm and zoom lenses, an electronic dimmer control for either 115 volts or battery, and a handy carrying case.

Arriflex will gladly demonstrate the Lightflex for any production personnel planning to use it. On their sample reel, Lightflex demonstrates how a rainbow can be added to an otherwise bald sky. They also show a room paneled in dark walnut and lit with only available window light. The walls appear almost black until the Lightflex is switched on, and the walnut panels come to life with a rich, warm wood tone!

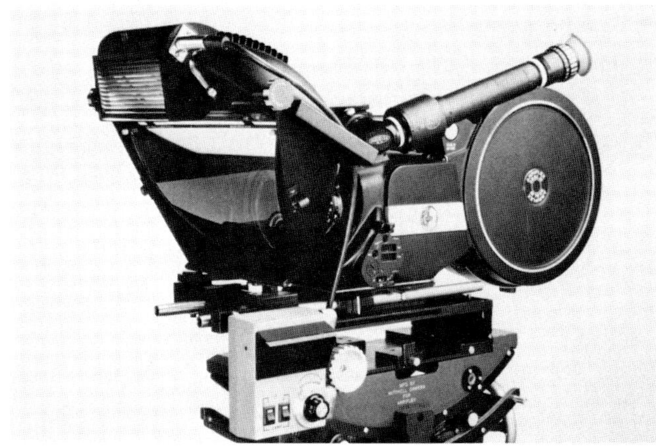

Figure 11 The Lightflex mounted for use on an Arriflex camera. The curved glass in front reflects light, color, or both from the overhead lightbox.

If the Lightflex can do all this, why isn't it used on every professional film production? There are currently several reasons for its limited use. It's relatively new, and when it first came out, the number of available units was limited. Also, all directors of photography work in their own individual styles. Some are such purists that they won't use anything in front of the lens except the #85 filter when filming in daylight; others prefer to use lamps to fill in shadows or gels to add color. Some cinematographers, unfamiliar with this equipment, may feel it takes too much time to install and use; however, considering the saving of some lighting and better picture quality, there's no time lost when using the Lightflex. There is no doubt as to what the Lightflex can do to add to the look of the picture. It has been used to great advantage on many well-received feature films.

The rental fees for the Lightflex are often offset by the time saved in lighting the set. By planning carefully, the cinematographer can rent the Lightflex for only those scenes in which it is worth its cost. In other words, it need not be rented for the entire production. Some testing must be done to learn the system and to be able to visualize the final effect on the screen, but once learned and used with confidence, the Lightflex is an ultimate image control.

Arriflex recently introduced a new item for cinematographers. Similar to the Lightflex in results, the Arri VariCon is a compact, contrast control unit that permits simple in-camera contrast reduction without affecting color temperature or normal skin tone. The VariCon allows simpler, more precise scene-to-scene contrast reduction than flashing. It consists of eight miniature halogen bulbs that surround the edges of a 6¼" special optical glass and cause it to glow. Light emanating from the glass brightens the film's darkest areas, lowering the image contrast. The output of the bulbs is controlled by a shutter, delivering constant color temperature at all settings. The VariCon also permits subtle coloring of the film's underexposed areas by means of internal dichroic filters, which deliver various colors of light to the film. At present the VariCon has virtually replaced the Lightflex. (See Figure 12.)

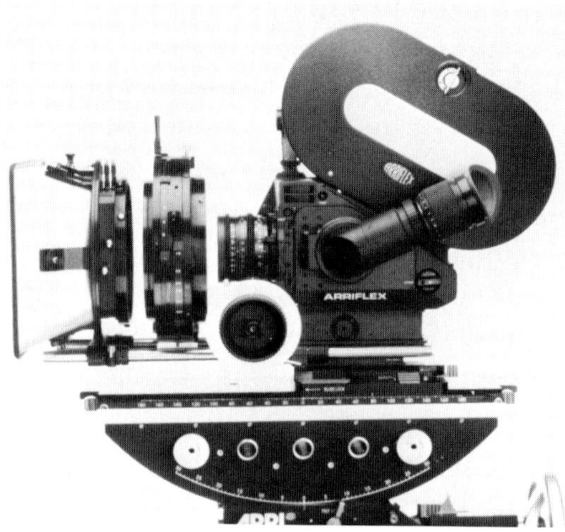

Figure 12 The new VariCon, made by Arriflex, performs like the Lightflex and will eventually replace it.

MAKING YOUR OWN CAMERA FILTERS

If your budget won't allow you to buy or rent a series of filters, don't despair! It's sometimes a simple matter to create your own.

I recently shot *The Neon Empire* on location, and although I had my usual large assortment of Panavision filters, an unexpected situation arose, and I wanted a special filter for a particular effect. I asked to see the colored gel assortment that the gaffer carries to use on lamps. I found a rose-colored Cinegel made by Rosco and cut a piece to fit a net frame, which in turn fit the Panavision matte box. I looked at the scene through the filter alone to judge its effect and to see if I could detect any distortion. It looked good, so I put it in the matte box and examined the image on the ground glass. It still looked good! Using a spot meter, I determined the filter factor of the Cinegel, made the F/stop adjustment, and filmed the whole sequence. When I saw the film the next day, the dailies supported my judgment. This particular sequence had its own special look, and the image was fine.

When making your own filters, you must first decide how the gel will be supported. You can run the gamut from just taping the gel to the front element, which is very temporary, or mounting it between two pieces of optical glass and taping the edges, which is more permanent. You can also make a frame out of anything sturdy that will fit your matte box and mount the gel on that.

The next step is to tailor the gel to its specific use. If it's to be used to change the color of the sky only, you will want it to conform to the angle of the horizon. The blend line—the edge of the filter that will be in the picture area—is an important consideration. A straight cut is the most difficult to hide. The saw-toothed edge created by pinking shears helps hide the edge of the gel. The best blend line is made by using scissors or a razor blade to carefully cut a deep saw-toothed edge with many points per inch. (See Color Plate 27 and Figure 13.)

Filters are most effective when the blend line is completely invisible. In addition to the shape of the edge of the filter, the blend line is also affected by the depth of field of the lens, which in turn is affected by the focal length of the lens, the lens stop used, and the proximity of the filter to the front element. Caution must be exercised when using a wide-angle lens at a relatively small lens stop. As the focal length increases and the lens stop is more open, the chance of seeing the edge of the filter is reduced. Always check the filter blend line on the ground glass with the lens stopped down, and if necessary, add a neutral density filter that enables you to film at a stop that's open enough to keep the edge of the filter invisible.

With an assortment of colored gels, there is almost no limit to the range of filters you can make. Gels can also be used in combination—that is, the upper half can be one color and the lower half another color—or a strip of color can be used in the middle of an otherwise empty gel frame. You can let your imagination run wild and create a very special look for each special scene. (See Color Plate 20.)

Rosco and Lee are excellent sources of quality colored gels. These manufacturers will send a complete gel sample book at your request. They're listed in the Appendix.

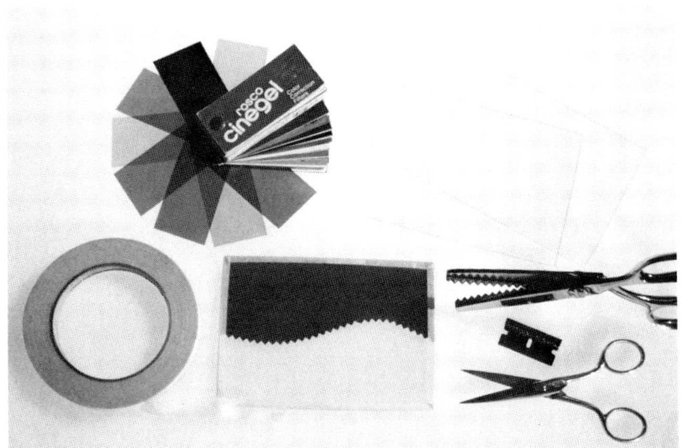

Figure 13 Shown are the items needed to make your own filters. Clockwise from upper left: colored gel sample book from which to order the desired color (sheets are 20" x 24") ; two thin pieces of clear glass in the required size; scissors, razor blade, or pinking shears; a completed homemade filter, a roll of one-half-inch masking tape to bind the pieces of glass.

CREATING A DAY-FOR-NIGHT EFFECT

You may have heard the expression, *shooting day-for-night* and wondered what it meant or, more probably, how to do it. To shoot day-for-night means to make a scene filmed in daylight look like it takes place at night. A good day-for-night effect is not difficult to achieve if all of the necessary elements are in place.

Let's assume you're using a color emulsion such as Eastman #5247 or a similar tungsten color negative. Of prime importance is the presence of clear sunlight in order to have both highlights and shadows. In a day-for-night scene, the sun in effect becomes the moon. Without black shadows, it's difficult to make a good day-for-night scene. Next, you'll need to position the camera to create a cross back light from the sun; in other words, the sunlight should strike your subject from an angle of about 90° to 120° from the lens position. This enables you to film the shadow areas with a highlight on one side of the subject. For wide landscape scenes, you must carefully pick the time of day to shoot; early morning and late afternoon usually offer the shadow and texture features that enhance the effect. Next, you must reduce the brightness of the sky to a point that is visually acceptable as night but not necessarily totally black. There are several ways to darken the sky. The simplest is to use a polarizer if you're shooting at about 90° from the sun and the sky is quite blue. If the polarizer doesn't make the sky dark enough, a graduated ND-to-clear filter can be used. If that doesn't work (because the action would cross into the ND portion of the filter, for example), try using a neutral blended ratio attenuator. If the sky is still too bright and may detract from the illusion of night, a higher camera angle that eliminates the sky might work.

If you shoot with normal exposure, the image will look like a dramatic day shot. Therefore, the next things to consider are color and exposure. It's best to print day-for-night negatives in the midrange of printer lights: this produces a good black density. Exposure should be reduced one and a half to two stops, depending on the scene. Underexposure eliminates most of the shadow detail and makes the sky darker. To add color and give the day-for-night moonlight effect, you have two choices: (1) remove the #85 filter, making the tungsten emulsion look blue when exposed to daylight, and compensate the exposure for the filter factor or (2) leave the filter on, and instruct the lab to print for night effect. The lab will reduce the red content of the printer light, creating the moonlight effect. I prefer the latter method because it offers more flexibility: the blueness of the "moonlight" can be adjusted. Another way to avoid a very blue moonlight is to substitute a coral #1 or #2 for the #85. The coral filter only partially corrects the tungsten film. It will still be blue but more believable. (See Color Plate 21.)

In the event you are using a daylight emulsion color negative, or slide, the technique for a day-for-night effect is slightly different. Because daylight emulsion has no #85 to remove, the bluish cast for the moonlight effect is created by adding a blue filter (EK#80A). This blue filter has a filter factor of 2x. After all your exposure compensations for the other filters that were required for a tungsten emulsion, the blue filter is added without further exposure compensation to create the necessary two stops of underexposure. Depending on the nature of the scene, you may want to underexpose only 1½ stops, in which case you'd partially compensate for the blue filter.

If you are working with a video camera and want to create a day-for-night effect, the technique is also slightly different. Video cameras have an internal white-balancing system. In order to create the blue cast necessary for a day-for-night effect, you must "fool" the white balance. Simply use the manual white balance adjustment to set the video camera for tungsten, which makes the daylight look blue to the camera. Then, follow all the above instructions for shooting on a tungsten-balanced color negative, and underexpose using the manual iris control. A monitor can help determine exactly how far to underexpose the scene.

The final ingredient for a good day-for-night effect is a source of light other than the "moonlight." In feature films, day-for-night scenes are usually relegated only to those areas that cannot be lit easily, such as expansive landscapes. City scenes that show stores, cars, street lights, and buildings are best filmed night-for-night, which means they are actually shot at night and lit. Shooting night-for-night is easy with the newer high-speed emulsions because less light is needed than before. A light source, such as auto headlights, house lights or flashlights used in a long shot in a day-for-night scene requires the source light to be much brighter than other highlights in the scene so that when underexposed the source light will still look very bright. If it's a car traveling along a road, extra-bright headlights replace normal ones, or small thousand-watt quartz lights, powered by batteries, are sometimes tied to the front of the car and pointed toward the camera to simulate headlights. In a long shot, the deception doesn't show. If the landscape has a house in the distance and you want the effect of a light in the window, there are two ways to accomplish that. The most common is to tape tracing paper to the inside of the window and put a very strong source of light directly behind it pointing toward the camera. Another way is to attach a sheet of 3M Company Scotchlite type 7610 to the inside of the window, facing the camera, put a reflector directly over the camera, and bounce the sunlight into the Scotchlite. Because Scotchlite reflects about 90% of the light that hits it, the window will be almost as bright to the camera as a sun reflector is to human eyes. If you want a warm glow from the Scotchlite, put a sheet of colored gel between it and the window, or use a gold reflector.

It's important to inform the lab of exactly how you want the scene to look. If the lab sees an unexplained underexposed scene, they may try to "save" it by printing on low printer lights, which will ruin the day-for-night effect. You must instruct the lab to print for "night effect" and tell them whether you had a #85 filter on the camera. Besides noting these on the camera report, it's also a good idea to make a phone call to the liaison person handling your film at the lab.

If you're careful about selecting the time of day, positioning the camera in relation to the sun, darkening the sky, and perhaps having a source of light within the frame, you should end up with a very good day-for-night scene. Make your final decisions about all of the elements of the scene by studying the image on the ground glass, especially the brightness of the light source.

The most practical advice on using camera filters comes from long experience on many theatrical feature productions. Many cinematographers use a wide variety of camera filters to create images that they cannot obtain in any other way. You must be the ultimate judge of how best to achieve the individual style that will give your film the best possible look for the story it tells.

Information about using filters would not be complete without describing this last dramatic effect: the removal of all filters! The absence of a filter can itself be an effect. Visualize one scene following another, all shot with filters, then suddenly on a cut—unadorned reality. The visual change can be a striking statement.

Chapter 6
Image Control in the Film Processing Laboratory

Image control is a fascinating aspect of cinematography. With the help of the film lab, you have further ability to alter the normal response of a color negative emulsion in a way that allows you to make a major contribution to the telling of the story on film. Up to this point, you've seen the controls that can be exercised in conjunction with the camera:

- selection of colors for the sets and wardrobe
- style of lighting
- choice of film stock
- selection of lenses and positioning of the camera relative to the sun
- choices of camera filters and lens attachments
- use of colored lamp gels or smoke for special effects

All of these contribute to image control, and all can be further enhanced through the contribution of the film lab. Some of these can also unwittingly be undone by the film lab! So your relationship and communication with the film lab are crucial.

THE FILM LAB

Although all film labs share certain obvious similarities, each film lab has unique characteristics. You need to ascertain which lab is the best one to become your partner in achieving your visual goals. This process of selection, like so many others in cinematography, begins with testing.

If you shoot the same scene under the same conditions on separate rolls of film of the same make and emulsion and send the exposed color negatives to different labs, you can compare the results on the screen. If three different labs were selected, there's an excellent chance you'll see three different images on the screen. One test may be much more contrasty than another, one may be more grainy, or one may lack color saturation. Don't worry about color differences from lab to lab because they are correctable, and a second print can prove that. However, too much contrast or grain may not be correctable, so it's important to know the difference. There's something else to watch for: if a plain wall of midtone values looks as if it's flickering slightly or appears to have a varying mottled effect, there may be a problem with density fluctuation in the negative due to the processing. Project the test a second time, get close to the screen, and see if there's a boiling effect in the grain structure of the image. It may be much more noticeable in one test than another, an indication that that particular lab's processing is contributing to the graininess of the image. Needless to say, be sure to identify your lab rolls so that you know which is which! Experienced cinematographers who are well aware of their own technique can tell almost immediately if a lab is doing a first-rate job of processing. They have no need to look at gamma charts or gray-scale lab tests; after all, what's on the screen is what counts, and that's what will be judged by the director, the producer, the audience, and the film critics.

To be completely fair with the film labs, you have to take into account that there can be slight variations in the color negative being processed. Film manufacturers do a meritorious job of making sensitive emulsions as consistent as possible in both the negative and the print raw stocks. However, glitches occasionally occur. When the lab gets samples of new emulsions of print raw stock, they make gamma and color tests before putting the recent arrival into the production line. They also test new negative emulsions and note the emulsion number to record any variant that shows up in their tests. They then know how to correct for a difference in color response, no matter how slight, when they later print that particular emulsion. As you might imagine, this is an exceedingly precise field in which to work.

When I first started as a darkroom assistant to a still photographer, he continually stressed that I must "respect" the material. Motion pictures even more than stills, offer an unparalleled opportunity to learn respect when you consider how many elements are necessary to create that huge, beautiful image on the screen. Don't forget that the origin of that huge, beautiful screen image is only about the size of a postage stamp!

With the advent of television and so much more film being exposed, the processing had to be sped up in order to turn out dailies for the next day. The industry advanced from normal negative processing in a 68° developer for 18 minutes to processing in a "hot" negative bath of 106° developer for only three minutes! This is how critical developing time is: a 15-second difference in the 18-minute developer is only an error of 1%, but the same 15-second difference in the "hot" bath is an error of 8%! Hopefully, a processing time error won't occur, but it sometimes does and this demonstrates how critical the processing is.

The film manufacturers supply the labs with recommended developing formulas and time and temperature charts for the processing baths. All film labs have basically the same types of developing tanks, although some might make their own modifications. Why then does the image quality vary from lab to lab?

Here are some of the possible answers. Any modification of the developing tanks can cause a difference in the processing standards. Developing tanks have spray jets that agitate the developer over the moving negative as it travels through the bath; if one jet gets clogged, a density fluctuation can occur. If the lab tries to send more than the usual amount of film through the bath within a given time (by increasing the developer temperature only slightly), the negative image will be affected. If any chemicals are improperly handled or if the replenishing rate isn't quite correct, it will affect the negative image. Many labs operate around the clock with three separate eight-hour shifts; if the personnel on duty don't oversee the work properly, "mysterious things" can happen that will affect the negative image. Any one of these circumstances can make a difference in image quality, and these differences can be seen early on in your tests.

You should also be aware that what appears to be a very slight defect in processing and is almost undiscernible in the dailies can become a major concern when release prints are made from a duplicate negative and that slight defect becomes much more visible. It's the cinematographer's responsibility to point out to the lab's quality-control person any deviation discovered between one day's dailies and the next. These very slight differences are sometimes difficult for the lab to detect because they usually run the dailies on high-speed projectors that can render a slight defect virtually invisible. I don't know of any lab that isn't glad to have information that lets them zero in on a possible problem.

The quality of the water used in processing is a variable over which neither the film manufacturers nor the film labs have any control. It's true that the labs have water filters and purifiers, but the water they start with does vary greatly. As mentioned in Chapter 1, the difference in water quality between the United States and Europe does affect both negative and positive images.

When I started in the film business, it was up to the director of photography to decide which lab would process the exposed negative and print the dailies. Depending on the look that I wanted—more or less contrast, more or less exposure latitude, or more or less grain—I could choose the lab that was best at producing these results. Today, the selection of a lab is often dictated by dollars instead of specific qualities. Industry managers are increasingly influenced by attorneys and accountants who are more experienced with multi-industrial corporations and mergers than with film production. In today's film industry, we see fewer filmmakers as heads of film production companies.

If you're aware of what can go wrong at the lab, you'll know what to look for and can perhaps get it corrected, or at least you'll know how to work with the problems of contrast, grain, and exposure latitude. For example, if the prints seem too contrasty, lower the contrast with filters or lighting while shooting; if the results seem grainy, additional exposure can help.

The frequency of lab errors is very slight. In many instances, film labs have saved the day for me, and they can save the reputation of any cinematographer. The film lab is truly your friend and should always be thought of as such.

FORCED DEVELOPMENT

One of the most popular lab techniques used to be forced, or pushed, development. Forced development, simply stated, means extending the time in the developing bath to give an apparent increase in the sensitivity of the negative, that is, increasing its ASA speed rating or its exposure index. With the advent of high-speed color negative emulsions, cinematographers today depend less on forced development, but it is still a valuable tool. Every now and then the director of photography may be faced with a lighting situation in which the shot can only be accomplished with forced development. This situation usually surfaces when filming under available light that is inadequate for even high-speed film.

I'd like you to hear from a long-standing friend of mine, now deceased, who was regarded as one of the foremost experts in the film laboratory processing industry—Sidney Solow. The following is from his contribution to the *American Cinematographer Manual,* edited by Charles G. Clarke, A.S.C. (Hollywood: A.S.C. Holding Corp., 1980, pp. 476-477).

> With the color films most commonly used today, it is possible to compensate for underexposure by extending development or "pushing." Similar to the principles of traditional black-and-white sensitometry, forced development of these color films increases their contrast, graininess and fog level. Therefore, forced development can never yield the same image quality possible when films are exposed and processed strictly according to the manufacturer's recommendations. In many instances, however, the image quality obtained with underexposure and overdevelopment is entirely satisfactory; and a cameraman may want to take advantage of this fact when shooting under adverse light conditions. What it means in effect is that he can deliberately underexpose his film (sometimes by as much as two stops) and request that the laboratory compensate in development.
>
> With Eastman color negative film #5247 and Fuji Type #8512 and Gevaert #682, the films most commonly used in 35mm professional production in the United States, it is possible to "push" one stop in development without appreciable loss in image quality.[1] The scenes produced by forced development can be intercut with scenes exposed and processed normally. If color negative is pushed two stops in development, the increase in graininess and fog level is substantial, but the results are acceptable for scenes involving night-for-night photography or available-light photography under exceptional circumstances. Extending development of #5247 beyond two stops does not contribute to the image any further, it only increases the grain and fog level and should not be attempted even as an emergency measure.
>
> It should be realized that with this type of film, especially, the sensitometric balance of the three emulsion layers is achieved only with normal processing and that forcing development does not accomplish a true compensation for underexposure. Forced development does not result in a substantial increase in Exposure Index of the

[1]*Author's note:* For interior photography or low-light conditions, these emulsions are rapidly being replaced by the newer high-speed Eastman #5296 and Fuji F-500 #8514, which often do not require forced development.

negative as measured by accepted scientific methods. Nevertheless, it cannot be denied that the technique proves to be of immense practical value by bringing the underexposed negative into an acceptable printing range.

Ektachrome Commercial Film Eastman #7252 (ECO), the reversal film used for most 16mm productions, can be pushed one stop without any appreciable loss in image quality. It cannot be pushed more than one stop, however, since the color balance becomes distorted. Ektachrome Video News Film #7240 (Tungsten) and Ektachrome Video News Film high speed #7250 (Tungsten), the reversal films most commonly used for news photography, can be pushed one stop with acceptable results.[2] In emergency situations they can be pushed up to three stops with some loss in quality. The ability to underexpose these films enables news and sports cameramen to capture moments on film which would otherwise be lost, but it should by no means be regarded as a suitable substitute for additional lighting when it can be provided.

If a cameraman anticipates the need for deliberate underexposure during a production, he should, if possible, shoot careful tests using the same emulsion that he will use and have them processed by the lab that will be processing the film for the production. The results can be analyzed with the help of the lab so that he can know what to expect when he underexposes scenes during production. Needless to say, underexposed rolls should be clearly marked with instructions as to how much they should be pushed when they are sent to the lab for development.

I would like to add that as a true laboratory technician, Mr. Solow believed that the ultimate color negative is achieved through following the manufacturer's recommendations for exposure and processing. However, he also states that "it is possible to push one stop in development without appreciable loss in image quality"(page 476). For what it's worth, especially for the uninitiated, I have filmed many feature films on both Eastman #5247 and Fuji #8512 and used forced development on all interior and night exterior scenes without the slightest noticeable effect on the image quality. However, to compensate for any increase in contrast, I sometimes used a low-contrast filter. Forced development of one stop enables you to use half the normal amount of set lighting or to stop down one extra stop. The latter is quite a convenience when filming with prime lenses intermixed with a zoom lens because the zoom lens is usually at least one stop slower. Another advantage is that the extra stop provides a greater depth of field, which certainly comes in handy when splitting focus.

After you've made your forced development tests with a particular lab, if you feel that, due to the processing, the contrast has increased more than you like, you can retain the advantages of the forced development and reduce the added contrast simply by using a low-contrast filter. It should be no problem to find the correct filter to compensate for an increase in contrast. You can test several low-contrast filters and pick the one that does the best job.

[2]*Author's note:* The use of videotape cameras for news coverage has almost eliminated the use of film.

FLASHING THE NEGATIVE

Cinematographers must sometimes film scenes that are naturally very contrasty. You might be able to solve the problem by adding sufficient lighting, by repainting the set, or by using the proper low-contrast filters. But imagine that you're filming a deep gorge or a mountainous landscape and the subject itself presents a great deal of contrast. There's an excellent lab technique to reduce the contrast: flashing the negative effectively reduces contrast when a low-level white light is used. Flashing with a low-level colored light can add an overall tint to a scene that is apparent only in the shadow areas. Once again I'll quote from the *American Cinematographer's Manual* (pp. 474–475) because the material is so concise and well written.

Flashing may be described qualitatively as subjecting negative film to a general fogging type of weak, controlled uniform exposure prior to development either before, during or after the act of running film through a camera to photograph the desired subject for reasons outlined below.

There is no measurable difference in effect if the flashing takes place before or after the principle exposure. As a result, because of various unfavorable factors such as not being able to control the time interval between the flash exposure and the time that development will actually take place, and not knowing the actual conditions of photography in advance, "pre-flashing" is generally avoided in favor of "post-flashing." Simultaneous flashing during actual photography by means of a special device attached to the front of the camera lens has been demonstrated successfully but has not attained any degree of popularity.[3]

Flashing has seemed to vary in popularity among cinematographers and seems to have become a controversial subject. Some directors of photography avoid flashing as a matter of principle, some use it only occasionally to accommodate specific situations, and some employ it to a greater or lesser degree on practically every picture they shoot. Among all these groups are cinematographers of great distinction, so no consensus regarding flashing is available.

Since color negative, as is well known, consists basically of three emulsion layers coated on a cellulose tri-acetate base, the spectral composition of the light used for flashing can be a neutral equivalent to tungsten light (3200°K) which would affect all three emulsion layers equally. This effect can also be achieved by the use of a chemical foggant in a weak solution just prior to the developing bath.[4] However, the advantages of using light to accomplish the flashing are several. The spectral composition of the light can be altered to distribute the flashing in different degrees among the three emulsion layers, the intensity of the light can be controlled conveniently and accurately for different rolls as requested by the cinematographer, and adjustment can be made to accommodate differences in emulsion batches if necessary.[5]

Unfortunately there is no uniformity among laboratories and cinematographers in designating the quantitative aspects of flashing or in relating degrees of flashing with effect desired. Cinematographers and laboratory men frequently specify certain "percentages" of flashing, but there are various interpretations of what these values mean. It seems that the most logical and scientifically acceptable definition of the degree of flashing is to use the increase in base and fog density of the blue layer after development. This increase would range in values from about .10 to as much as .50. For those who prefer to express degree of flashing in terms of percentages, a .20 increase in density of the blue sensitive layer would be expressed as a 20% flash.[6]

The fundamental reason for using this type of flashing is to reduce the contrast of the image. This effect is accomplished because the flashing exposure affects solely the shadow densities which are, of course, in the region of the toe of the sensitometric curve of the film.[7]

Occasionally, a cinematographer may be confronted by conditions where he cannot have access to lenses of uniform contrast and the necessary contrast control filters and where he must call upon the laboratory to subject scenes which would otherwise be excessive in contrast

[3] *Author's note:* Although not in popular use when the above was written, the Lightflex device can create an effect similar to flashing. See Chapter 5 under "Lightflex."

[4] *Author's note:* See "Chemical Fog" in Glossary for explanation of Chemtone.

[5] *Author's note:* Different emulsion batches used to vary quite a bit in their color balance. Now, however, the controls are much finer and only very slight color differences may appear.

[6] *Author's note*: The normal request for flashing is in the range of 15% to 30%.

[7] *Author's note:* Flashing the negative results in an appreciable increase in the shadow detail without any affect on the highlight areas.

to various degrees of flashing in order to equalize the photographic effect of the various lenses.[8]

Another reason for flashing is to achieve certain creative, artistic effects by using a non-neutral flashing exposure which would then alter the normal color rendition of the developed negative. It is a matter of continuing controversy among cinematographers as to whether distortion of the basic color rendition is necessary or wise. A good many cinematographers have relied upon laboratories to achieve any desired creative color moods on the ground that there is infinite variation possible during the post-production phase, whereas shooting a non-standard negative may result in being locked into a particular interpretation.[9]

Still another use for flashing is to help achieve a higher effective emulsion speed in connection with forced development. Since force development tends to raise the contrast, flashing tends to compensate for that sometimes undesirable quality. Thus the combination of both actually results in increasing effective speed without appreciably detracting from the photographic quality.

Another important use of flashing is to achieve matching contrast between different films. This use is particularly valuable in the field of 16mm reversal films where the most commonly used material is Eastman Ektachrome Commercial Type #7252. This material has an exposure index to tungsten of 25 with the ability to be pushed without penalty to EI 50. For greater sensitivity for certain low light level situations, the cinematographer must resort to Eastman Type #7241, #7242, #7240, and #7250. All of these are higher contrast films and intended to produce appropriate contrast for direct projection. Without flashing, their contrast is totally inappropriate for photographic printing. As a result, when scenes on these films are to be incorporated with scenes photographed on ECO film, it may be considered universal practice to flash the high contrast films to the extent that their sensitometric curves are brought into an almost perfect match to that of ECO.

This extract tells the story completely, but let me add something from my own experience. During a time when emulsions were being changed by Eastman, I was given two rolls of film: one was a discontinued emulsion that I knew was more contrasty, and the other was a new, improved emulsion. I had no choice but to go ahead and shoot. I asked the lab to flash the older roll 20%. When it was intercut with the newer, less contrasty emulsion, no difference in contrast was discernible.

Flashing the negative is another tool for cinematographers to consider in the overall design of the film.

FLASHING THE PRINT

Flashing a reel of print stock before printing exposes it to a small, controlled amount of white light. This technique is used to reduce the tonal range and thus the contrast, but it also gives a different look to the release print. The results of flashing the print are seen in the highlights rather than the shadows; this effect is the opposite of flashing the negative. Flashing the print reduces the brightness of the highlight areas and thereby reduces the tonal range. When you are confronted by extreme contrast in a scene and want greater shadow detail, you can increase the exposure to favor the shadows, and the resulting additional brightness of the

[8]*Author's note:* Such scenes must be filmed on separate film rolls, and flashing instructions must be given to the lab.

[9]*Author's note:* It's exactly for this reason that I would rather have the desired effect in my original negative: I'd rather that it be my interpretation and not someone else's. Also, flashing with a non-neutral exposure that affects the shadows more than the highlights imparts a different look than that obtained with a color filter, which would also affect the midrange and highlight areas. In post-production, cinematographers are often not available to supervise the desired effects. During post, if the lab is to use flashing, the final negative to be used for that particular scene or sequence will be a dupe negative, not the original. If only the shadow areas are to be affected, as in the case of flashing, a similar effect can be created on the original color negative with the Lightflex or VariCon.

highlights can be reduced later by flashing the print.

You must try different flashing percentages and base your decision on the visual effects created. With flashing a negative, a particular scene or group of scenes can be flashed by separating those scenes on a separate roll of film; however, when flashing a release print, the entire thousand-foot reel must be flashed. This is a very important consideration and requires careful planning. Consider also that every reel of the release print that requires flashing must be flashed individually. I don't mean to suggest that flashing the print is too painstaking a process to use; on the contrary, it has been used very effectively by many experienced directors of photography.

ENR PRINT PROCESSING

A printing process perfected by Technicolor Laboratory, ENR processing (named for Ernesto Navello, the technician at Technicolor Rome who created the technique), has a very unique and spectacular effect. ENR processing is done only during release printing. The effect is a distinct increase in the black density of the print and greater color saturation. The process excels in creating black blacks for night exterior scenes or low-key interiors where the black areas often appear smoky or slightly grayish, possibly due to underexposure and the subsequent lower printer light. The process also provides a moderate increase in the visual contrast of the scene due to the more saturated colors. A comparison of the same scene printed normally and printed with the ENR process will show an increased color saturation overall and a starker image.

In this process, the negative develops for a predetermined length of time in a black-and-white developer immediately after the color developer. This procedure enhances the silver halide particles in the emulsion that were not affected or were only partially affected in the color developer bath. Because this process is applied only to release prints, scene-by-scene effects cannot be done; a full thousand-foot reel must be processed at one time. All special processing involves extra cost, but it's certainly worth the expense if it recognizably supplies the special look you wanted.

PULL PROCESSING

Rarely used but available in some labs, pull processing is a method of developing in which the time in the color negative bath is reduced by a predetermined length. The results are the opposite of forced development. For pull processing to work effectively, the color negative must be overexposed one full stop and then underdeveloped. The resulting image has reduced graininess because greater proportions of the slower speed, finer grains of silver halide were exposed. There's also a moderate reduction in visual contrast, and the color balance of the negative remains unchanged. The results are excellent.

This type of processing is limited by the fastest speed at which the developing machines can operate, and not all labs are able to supply this service. Pull processing is primarily requested when the cinematographer wants a negative with as fine grain as possible for future optical effects or to counteract the contrasty processing of a particular lab. Like any other variable of color negative processing, the director of photography is well advised to make sufficient tests. Eastman has recently produced a finer grain high-speed emulsion, Eastman #5295, specifically for optical work from an original color negative.

CHEMICAL FOG TREATMENT

At one time, another way to control contrast and achieve greater shadow detail was through a proprietary process called *Chemtone,* which was used at TVC Laboratories, Inc. in New York. The process, although discontinued, is still interesting.

In the Chemtone process, a chemical foggant was added to the developing bath. As the chemical fog level was increased, it added sufficient density to the thin densities of the negative, represented by the toe of the sensitometric curve, and enabled them to record detail otherwise lost in the shadow areas. This process also reduced the tonal range somewhat, thereby lowering the contrast ratio. From personal experience, I can relate that the procedure worked quite successfully. Chemtone was very popular for a long time. However, with the advent of the low-contrast print stock for television release prints, the more universal use of filters to control contrast, and other lab processes used to gain shadow detail, the Chemtone process was discontinued.

TIMING THE ANSWER PRINT

After all the preparation for a feature film, after all the work involved in the actual production, after all the editing of that film, and after all the optical effects, the director of photography's final contribution is to collaborate on the final timing of the film. The process of timing a film ensures that all the scenes of a sequence have the same correct color balance and light levels. It's during the timing that a scene can be made lighter or darker, and the color can be shifted, if necessary, to make scenes match that were filmed on different days under varying light conditions. The process is usually repeated several times until the final timed answer print has a continuity of color and brightness throughout.

Timing a feature film requires the concentration of both the director of photography and the lab timer, a highly skilled technician. The timer's expertise puts the finishing touches on film production, and in recognition of this fact, he or she receives screen credit on the release prints.

Possibly one of the most shocking experiences for a director of photography is the screening of the first answer print! The answer print is the first complete print that incorporates all of the edited footage, any optical effects, and the titles. The shock comes from seeing scenes that looked fine during the dailies but that now seem to have great mismatches in color and density because they've been edited together. You see footage from the first day adjacent to a scene filmed on the fortieth day and they don't match at all! This is nothing to worry about; it just seems peculiar that the eye is so accommodating that you never noticed those discrepancies before. Labs are occasionally asked to make timed dailies, which look much better when edited because many of the color and density corrections have already been made. However, timed dailies are seldom made because of the increased cost.

When timing a print, the director of photography views the screen while sitting next to the timer, who makes notes on the density and color of each scene. I prefer not to sit with the timer until after the first two answer prints have been made and all the obvious mismatches have already been corrected by the timer. The colors that the timer deals with are cyan, magenta, and yellow. The densities of these colors are referred to as *printer lights*, or *points*, and eight points equal one stop of density. The overall range of printer lights is 1 to 50, but the lab likes to keep them between 25 and 40. The subtleties of timing require a keen eye to detect whether a point or two of magenta should be added or perhaps three points of yellow taken out to have two adjacent scenes match perfectly. If the color seems correct on the print but the scene is too bright, the timer may suggest that the print be made four points darker "across the board," that is, all colors be made equally darker. The four points are equal to one-half stop.

The lab timer often requests a copy of the script to better understand the dramatics of the story and govern the print densities accordingly. The director of photography offers suggestions to help preserve the nuances originally discussed with the director regarding the look of the film. The mood of a particular sequence can be drastically changed just by printing the film lighter or darker! Also, the effect of the camera filters can be altered by lighter or darker prints, so the timing of the final answer print is of utmost importance. Although the timer would turn out a good print without the director of photography's assistance and the editor and director would be on hand to give their approval, I believe that the director of photography must try to be present when the final answer print is timed. The special choices and contributions that the director of photography made during the actual filming must be escorted to completion or a whole layer of artistic intention can be erased. When I know that I won't be available for the final timing, I always try to put the effect I want on the negative when shooting; I do not leave it to the timer or anyone else to add later. All the notes made on camera reports at the time of filming, guiding the lab toward the desired look, have long since been filed in the editor's room, yet directors of photography always seem to remember the details of how they wanted a scene to look! Without any notes at all, they're able to express just how a scene should be printed.

This chapter should have made you more aware of the many additional controls extended to you, courtesy of the film lab. As mentioned earlier, the film lab should be considered a friend. Your relationship can also be seen as symbiotic: the lab wouldn't exist without your film to process, and your job wouldn't exist if there were no lab to process the film. Although you may not need any special lab processes now, the special methods of color negative processing may someday help you give your film an individual and memorable look.

Chapter 7
Considerations for Post Production

When involved in the planning stage of a new film, cinematographers must consider post-production work in relation to the techniques they may use during shooting that will alter the otherwise normal image. What exactly is post production? As the name implies (it's often called *post* for short), it's any work done on the film after the principal (main body of) production is completed: the addition of shots that were skipped and then rescheduled, inserts, or stock shots; the dubbing of dialogue, scoring music, or Foley effects; editing; titles; optical effects; and finally release printing.

OPTICALS

While shooting principal photography, the cinematographer must be concerned about any post-production work to be done to any of the scenes. For instance, if certain sequences call for extensive optical work in post, the cinematographer must be cautious in the use of diffusion, fog, low-contrast, or color filters. If a sequence calls for matte work of any kind—that is, blue screen, stationary art matte, or other composite work—it's best not to use any filtration on the original negative. Diffusion can sabotage efforts to get a sharp matte edge. Fog or low-contrast filters used on the original negative can cause a mismatch with post composite elements that do not have these filter effects. Shoot the original scene "clean"—that is, without filtration—combine it with the matte shot in post, then add any final optical effects to this newly created composite scene. The optical department can make the necessary adjustments (adding fog, changing the contrast) to bring this composite scene into balance with other scenes in the same sequence. This is much safer than filtering the original negative.

DUPE NEGATIVES, INTERPOSITIVES, AND INTERNEGATIVES

Another consideration for the cinematographer is that release prints of the motion picture are made from a duplicate, not the original, negative. When hundreds of release prints are made, the dupe negative serves to protect the original negative in the event of damage. In the process of making a duplicate color negative, there's often a slight increase in the contrast, particularly with the color reversal internegative (CRI) process and sometimes even with the interpositive/internegative (IP/IN) process. With excellent lab work, it's possible to create a dupe negative that produces a print that will compare favorably with a print from the original negative.

Effects made on the original negative offer better quality prints than effects made during post production. As an example, a simple vignette effect done in the camera with nets or a vignetting filter is an integral part of the original color negative. If it's done during post production, the effect is made optically using the original negative to create a duplicate negative that incorporates the vignette effect. This dupe is spliced in with original negative scenes and duplicated again for the release print. At this point, the scene that required the post-production vignette is a third-generation negative because it's been duped twice!

This is reason enough to do as many image-control effects as you can on the original negative. Of course, if the exact effect or its place in the story cannot be pinpointed during production, you have no choice but to wait and let post-production personnel take care of it. If intricate optical effects are scripted, they must obviously be made during post production. If you want an overall color change only in selected scenes—for example, a warm sunset, firelight, or candlelight—this can be handled by the lab. It's sometimes better to have the lab do these warm tone effects than to do them in the camera because a filter such as a heavy coral can filter out too much of the blue sensitive layer that is necessary to get a good black density in the print, especially in low-key scenes.

If possible, try to make yourself available for consultation during post production as well as during the final timing of the film. It's during the timing process that all final image changes are made, assuming that all the optical work has been critically assessed for quality and correctness. It's not uncommon to need three, four, or more color timing sessions to produce the conclusive, color-corrected print.

FILM-TO-TAPE TRANSFERS

Film-to-tape transfers are made during the post-production period if the film is destined for television or videocassette release. If you know beforehand that the film will be used only for television, you may want to reduce the overall contrast somewhat to improve the quality of the televised picture. Remember, film can handle a tonal range of about 140:1, but television can handle only about 40:1. Therefore, to help create normal-looking scenes, you can use a bit more fill light as long as you don't spoil the desired effect. Night exteriors and low-key scenes must be carefully and specifically lighted for good television transmission if you want detail in the shadow areas.

It's very informative for filmmakers to attend film-to-tape transfer sessions to see the degree of control that's available at the transfer console. The push of a button or the turn of a dial can increase or decrease the brightness or the contrast of an image; shifting the color balance is also a simple matter. In fact, much greater control of the image is possible at a film-to-tape transfer session than in the film lab. The wonderful thing is that no guesswork is involved because you're able to see the results immediately on a video monitor.

When making film-to-tape transfers, the tele-cine operator can benefit from the director of photography's input to produce the television image that most closely matches the original print. For a long time, the video operators who made film-to-tape transfers relied heavily on what their meters and oscilloscopes told them if they didn't have input from the director of photography. Many of my feature films that were transferred to tape have been diminished by being made much lighter than the original. They looked good on the "scopes" but not so good on the television tube!

There are currently two ways to make film-to-tape transfers. In the first method, the timer at the lab makes a first answer print from the edited negative after getting information from the director of photography regarding the desired look of the film. The timer makes further corrections of the density and scene-by-scene color balance. A second answer print is then made, and the timer notes any new timing changes for each cut of the film. The next step is to take these final timings and use them to produce a low-contrast color print. This special print is made on Eastman low-contrast positive print stock (#5380 for 35mm and #7380 for 16mm). An interpositive is sometimes made for a film-to-tape transfer because it too is a low-contrast print.

This low-contrast print need not be projected because it does not represent the ultimate image. During the transfer to tape, additional subtle changes in color and density can be made. A wet (liquid) gate is used for the print to get the cleanest image. The end result is a one-inch video master tape.

Video technology changes and improves rapidly. State-of-the-art video editing and transfer processes now use digital video equipment. This is the second and most cost-effective way to go from film to tape.

During shooting, an electronic slate using a continuous Society of Motion Picture and Television Engineers (SMPTE) code identifies the scenes and takes for the editor and the tape staff. Below the code on the same slate, traditional scene and take numbers are written so that the filmed material can be located in the script and good (OK) printed takes can be distinguished from "no good" (NG) takes.

The original color negative is used to transfer the OK takes to one-inch tape; no film print is necessary. The digital system uses the SMPTE code to search for and find these OK takes. Using information from the camera reports and the director of photography, if present, the video technician can make adjustments to improve the look and mood of the scenes. A half-inch dub (copy) is made at the same time. It serves as dailies for the director to choose the preferred takes with which the editor will edit the film.

The entire editing process is done electronically. The digital video equipment selects all the scenes by referring to the SMPTE code and assembles them in edited order. This edited footage is again transferred (only the second transfer), and any final corrections are made now that all scenes are in their final positions. During this transfer, the director of photography's input is invaluable. In addition to its speed and cost-effectiveness, one of digital video's great advantages is that there is no loss of image quality regardless of the number of transfers or the number of copies made, and the negative need never be cut!

In conclusion, recognizing the ultimate use of the film can be very helpful. However, if you're making a film for theatrical release that may later be released as a videocassette, you cannot be concerned about this eventuality. You must create the screen image that's right for the theaters. (This is not the same as planning ahead for optical work.) I was once told by the producer of a film I was about to shoot that the film would be shown in drive-in theaters, as well as indoor theaters, so would I please make my photography bright because drive-ins notoriously have dark screens! Another time, I complained that a studio screening room used for dailies had too dark a screen, a fact confirmed by my meter reading of the screen. The head of the camera department suggested that I add more light to my scenes so they would project better!

Chapter 8
The Definitive Lab Test

The importance of lab tests has been mentioned throughout this book, but you may not have had a full appreciation of just how crucial the information they offer is until now. When you consider all the techniques discussed thus far and their possible combinations, you can understand how carefully you must proceed to get everything just right.

A definitive lab test is needed so that you can evaluate exactly the quality of a color negative produced by a particular film lab. The test should convey the following information:

- The negative density that's optimum for that particular lab. This establishes a basic ASA exposure rating, or exposure index.
- The latitude of the negative emulsion in relation to that lab's processing, that is, how much the negative can be overexposed or underexposed and still render a good print.
- The graininess of the negative developed in that lab and the effect upon graininess from overdevelopment or under development (forced or pull processing). For greatest validity, a second identical test must be shot and sent to a different lab or labs so that you can compare the results.
- A comparison between normal and forced development (pushing one or two stops), if you plan to use this lab technique.
- A check on the general quality of the lab to uncover any density fluctuation or other problems caused by improper processing of the negative or print.
- A check on the contrast inherent in the negative when processed and printed by the first lab as compared to a test sent to other labs.

Once these characteristics have been established and the cinematographer understands how a particular lab works, he or she can determine how to alter the image to produce a print that represents the desired look of the film. As a result of the lab tests, the cinematographer may recommend that a specific lab be used. Under many present-day contracts, though, the director of photography has absolutely no voice in selecting the lab for a particular production. The producer or studio head may already have an existing arrangement with a lab, which is often based on monetary considerations and may not have any bearing on the quality of processing.

SETTING UP THE LAB TEST

If you have a large number of exterior scenes to film, you may want to do only exterior testing; however, an interior test best exemplifies the several steps to take in preparation for the test. Choose a place with a blank wall. Cover one section with a medium-gray poster card (about 3' × 4', available from art supply stores), another section with a white card, and the third section with a flat-black card. Have the model sit about eight feet in front of the wall next to a century stand (a stand with adjustable risers and arms to hold a variety of devices) that will hold an information card of about 12" × 16". It's premature for any kind of effect lighting, so just light the subject normally with a standard highlight-to-shadow ratio. Set the exposure meter to the manufacturer's EI or ASA rating. Use only one lens to eliminate mismatched lenses as a test factor. You need to have a range of two to three stops overexposure and two to three stops underexposure. Lighting for a stop of F/5.6 usually suffices because three stops overexposure requires a lens of F/2.0, which is generally available.

Frame the scene to include your model in medium close-up, the information card with color chart and gray scale, and include the three poster cards in the background. Check that the light on the background is balanced: the gray card should look normal compared to the flesh tone, not lighter or darker than its true color. Check also that the lens is protected from any back lights. Now you're ready to write the pertinent information on the 12" × 16" card:

- emulsion (type and number)
- exposure index
- processing (normal or forced)
- foot candles (equal to F/5.6)
- lens stop
- exposure (normal, one-half stop over or under, and so on)
- lab (especially if there's more than one)

This card should stay in the frame all the time, so that when you look at the printed test, you don't have to refer to a separate piece of paper for the relevant data. If you keep film clips of the test for future reference, all the corresponding information will be right there within each frame. (See Color Plates 22 and 23.)

SHOOTING THE LAB TEST

It's important that the lighting remain exactly the same throughout all parts of the test. You need to shoot about 30 feet (20 seconds of 35mm) for each part of the test, so there will be enough time to study all the details on the screen. This length also allows all the takes to fit on one four-hundred foot roll. Shoot the first take at F/5.6 (normal), then overexpose in half stops (F/4.5, F/4, F/3.2, F/2.8; F/2.3, F/2). When you change the lens to F/4.5 also change the information on the card to F/4.5 and change the exposure entry from normal to one-half stop over. Cover the lens, and run a few frames of blank film to separate the takes. Continue to change the F/stops until you've overexposed by two or more stops. After completing the overexposure section of the test, make another normal exposure at F/5.6, then shoot the underexposure section in half-stop increments of F/6.3, F/8, F/9, F/11, F/12.5, F/16, changing the information card each time.

If you're making a comparison test for another lab, reload the camera, and repeat the above test exactly. Don't forget to add the name of the lab to the information card as well as the film can so you'll know which roll of film is for which lab! If you are testing the forced development of one stop to compare to normal development, reload, change the card to read force one stop, and add a .30 neutral density filter. Using a ND filter for exposure control is better than stopping down an extra stop for each take, because it ensures that the lens quality remains the same at each stop. If you don't have a .30ND available, adjust the lens stops to compensate for the forced development. Take care that the camera reports with the instructions for the lab match what you actually did in your tests. Any special lighting tests should be made separately after evaluating the lab with the results of the first test. Do not use colored filters on the first lab tests. Unless the test is for a film production that will be shot entirely outdoors in daylight, it's best to test in an interior set where you have more control. Outdoor lighting can vary during the testing period, so exercise caution when shooting exterior tests.

INSTRUCTING THE LAB FOR THE TEST

The instructions to the lab should include the number of the raw stock emulsion, the type of processing desired (normal, forced, pull) and instructions to make two prints of everything. Make it clear that the first print is to be printed on the lab's normal "middle-of-the-scale" printer lights, without regard to the density of the negative, and that the entire roll be printed on that same printer light! This is called a *one-light print.* Also instruct the lab to print the second print for the best possible flesh tones at each exposure on the test roll. Why obtain two different types of prints? Each will give you different information.

ANALYZING THE LAB TEST

The first print indicates whether you should use the manufacturer's recommended exposure index with that particular lab. If the print of the take marked "normal" (in this case, the one shot at F/5.6) does in fact look normal as far as flesh tones are concerned, use that film speed as the basis for any normal filming or as the basis for determining any exposure change for "effect" filming. However, if the take that was underexposed one-half stop, photographed at F/6.3, has a better skin tone and richer color, then with that particular lab's processing, you can safely rerate the exposure index to a higher number. For instance, if the manufacturer's recommended film speed has an EI of 320 and you see that the take underexposed one-half stop (printed on the lab's middle-of-the-scale printer light) looks best, then rerate the film speed to an exposure index of 500, which is one-half stop faster than ASA 320. If you feel that the half-stop underexposure is a bit too dark but that the normal was slightly too bright, split the difference and rerate to an exposure index of ASA 400.

This first test will also indicate how much you might want to underexpose a scene for a low-key night interior. Because everything was printed on the middle scale of the printer lights, even one full stop of underexposure will still be of sufficient density to render an excellent print and to produce an excellent solid black color. (See Color Plate 23.)

Screen this first test again, and concentrate on the gray card in the background. Look at it very closely to determine whether it appears alternately a bit lighter and darker or whether there's a slight flickering or wavering effect. This usually indicates density fluctuation, which always shows up most prominently in the middle-gray values. Density fluctuation can be caused by a faulty printer, meaning that the original negative is fine. (The only way to be certain is to have the negative reprinted on a different printing machine. Lab personnel can tell from their records which printer they used for your tests.) If the density fluctuation can also be seen on the other print, ask a lab representive to sit in with you in the screening room to observe the problem and suggest a cause. Density fluctuation is often caused by a clogged jet or two in the developing tanks. If you can discern no similar density fluctuation in a test from a second lab, it might be advisable to use that second lab, or to at least get assurances that the problem will be solved in the first lab. Although density fluctuation may be very slight in the original negative and the producer may not be concerned about it, any post-production work—particularly when a dupe negative is made for release printing—will magnify the problem. Be extremely critical when viewing for density fluctuation.

Screen the first test a third time, get very close to the projection screen, and concentrate on the grain of the color negative that shows up on the print. Because the positive print stock has a fine grain structure, the grain you see is that of the negative. A grainy negative appears to be "boiling" and the effect is most easily seen in a medium-tone area of the picture. Concentrate on the gray card because very light or dark areas do not show the grain as easily. This boiling effect is merely the grain particles as they appear in different spots from frame to frame. Try to memorize what the grain looks like and then screen a test from another lab and see if any difference appears. You can also compare the grain of a normally processed negative and one that's been force-developed. If the lab that did the forced development seems to create a grain problem, you may want to reconsider using that procedure. Because forced development is a deviation from the norm, the effect on the grain varies from lab to lab. Generally, however, the grain increases only very slightly—not enough to cause problems—but it is important to know what the process will do.

With the advent of the new high-speed emulsions (Eastman #5296, Fuji AX #8514, Agfa Gevaert #XT-320), the need to force-develop is greatly reduced. The scenes that still require forced development when using minimal or available light are often expansive night exteriors that contain large black areas that don't show grain; thus any increase in grain size is not apparent.

Next, look at the second test print. If the lab has succeeded in getting the best possible print from each of the different exposures, the first thing you'll observe is that there's a great latitude of exposure in the color negative emulsion. This simply means that any moderate errors in exposure (either over or under) can still be printed as a good screen image. However, you can learn much more. You will probably see that almost all of the overexposed takes print very well, although when you've overexposed more than two stops, you might notice slight color distortions in the very pale colors or even in the flesh tones. However, on the underexposed tests, you will see by examining the black background card that when underexposed more than one and a half to two stops, the image seems to deteriorate more rapidly in color and it becomes more difficult to obtain a good black. The flesh tones become muddy, and the blacks start to go gray, smoky or greenish gray. Remember that a good black is a product of the printer lights, so if the lab uses printer lights that are too low in order to make a badly underexposed scene look normal, you'll soon lose the strong blacks of the scene. Thus, there's more leeway for error on the overexposure side than on the underexposure side. (See Color Plate 24.)

You can learn more from the second test. If you compare the gray card area of a take underexposed by one stop to a take overexposed by one stop, you will see that overexposure results in a reduction of grain.

The white card will show you whether the lab used correct printer lights for a white balance. If they didn't, that card will take on a color cast. The white card also helps show any color shift in the more overexposed takes.

Be careful not to confuse the second test with the first regarding underexposure for special effects. If you underexpose for low-key or night scenes by one-half to one stop, you still want the scene to print in the midrange for a good black, letting the skin tones go darker than normal for the effect. The second test will show that if you underexpose by one stop for additional film speed and want a normal print, not an effect scene, the lab is printing on the low edge of acceptable printer lights. Because there's no leeway for error or adjustment, you may lose the black tones of the scene. Again, a good solid black in a print is contingent on the printer lights, not the exposure. If the printer lights are too low, you cannot get a good black.

This is a good time to review the subject of printer lights which was discussed in "Exposure" in Chapter 3. There are fifty printer lights on the color negative printer. Printer light 1 is the lowest level of brightness, and 50 is the highest. There are eight printer lights for each stop of exposure. In other words, if you were to accidentally overexpose by one stop, the lab would adjust the printer lights to eight points higher and produce a perfectly good print.

If you decide to use an exposure index higher than the recommended one—to use less light or to stop down for greater depth of field—and the result is one stop of underexposure, there's little to worry about because a timing adjustment can be made in the printer lights with perfectly good results. Some cinematographers prefer to underexpose slightly in accordance with their style, and questionable results rarely appear on the screen. Other directors of photography with post-production work in mind, deliberately set their exposure index lower to overexpose. They get a denser negative with less grain, which prints in the high 30s to low 40s on the printer light scale. The choice is yours, but it's always a good idea to leave some leeway so the lab can print up or down, as necessary.

Evaluating the grain of the image is part of the analysis process, too. Negative emulsion is made up of two sizes of silver halide particles. The larger particles are more sensitive to light than the smaller ones. When you underexpose and then force-develop, the larger particles, or grains, are more fully developed than the smaller ones, and a grainier negative results. Conversely, a slightly overexposed color negative shows less grain than even a normally exposed negative, because more of the smaller, less-sensitive fine grains are exposed and developed. (See "Pull Processing" in Chapter 6.)

Although they might cause a slight increase in grain, cinematographers should not be discouraged from creating good low-key shots, either interior or night exterior. However, if much post-production work will be done with your negative, grain can be a consideration. New Eastman emulsion #5296 and the Fuji F series emulsions incorporate a new structure known as *T-grains,* which were designed for special application in blue screen matte work and other optical effects. Inherent in these emulsions is a finer grain.

After you have fully evaluated the lab tests, decided on an exposure index, and learned how the lab treats the negative in regard to contrast and grain, you can then make plans to shoot further tests for the look you want for your film. This is the time to try lighting effects and any unfamiliar filters. The testing procedure is time consuming, but the information it provides is invaluable for adding to your experience and confidence.

THE SCREENING ROOM

Be sure to view the tests in the screening room at the same lab that will time the prints. Laboratory screening rooms have the correct degree of screen brightness, and they provide an even level of illumination across the screen. Without these screening conditions, your tests can be invalid. It's a sad truth that the screens in many projection rooms, even at major studios, have the wrong intensity of illumination. Too little light or uneven light across the screen makes it impossible to accurately assess a lab test, let alone the production dailies. If the screening room assigned for your dailies has poor screen brightness or uneven lighting, it's up to you and your producer to pressure the facility to remedy the situation.

The reflected brightness of a screen is measured in foot-lamberts (ft-L), the standard being 16 ft-L at screen center, with not more than a 20% fall-off at the edges. You can check the screen accurately with a spot meter specially calibrated to read foot-lamberts; the projectionist should have one you can use. If not, you can use your foot-candle meter to take an incident light reading, pointing the meter towards the projector. Both types of readings should be taken with the projector running without film in the gate. The foot-candle meter should read approximately 12 to 14 fc. Note: Foot-lamberts measure reflected light, and foot-candles measure incident light; there is no correlation between the two.

THE LENS TEST

Another test that should be made before starting a production and that may have some bearing on filter selection is a lens comparison test. With each of the lenses that you plan to use, film a test chart to indicate not only the focus, calibration, and abberations of each but also to compare contrast and color reproduction. The results of this test sometimes indicate that a particular lens has significantly more or less contrast or shows a color difference. These are most noticeable when comparing a zoom lens to several prime lenses. There is no need to worry about color differences: they can easily be corrected in the final timing of the film. If a lens is more contrasty than the other lenses, you can do two things: exchange the lens for another of the same focal length or find the low-contrast filter that will make this lens match the others you plan to use. If a lens is found to lack contrast, you must exchange it; fortunately, this rarely happens. If you want all the dailies to have color consistency, a gelatin color-compensating filter can be attached to the rear element of the off-color lens, thus bringing it into color balance with the other lenses. This is important when working with 16mm reversal color films. Because there is no color adjustment in the lab when using reversal film, there would be limited color adjustment when duping the original. A gel filter on the rear element is well protected, less likely to be forgotten than a glass filter, and avoids adding a second or third glass filter in front of the lens. Experienced camera assistants know how to shoot valid lens tests using a focus chart, a color scale for color reproduction, and a gray scale for contrast compari-

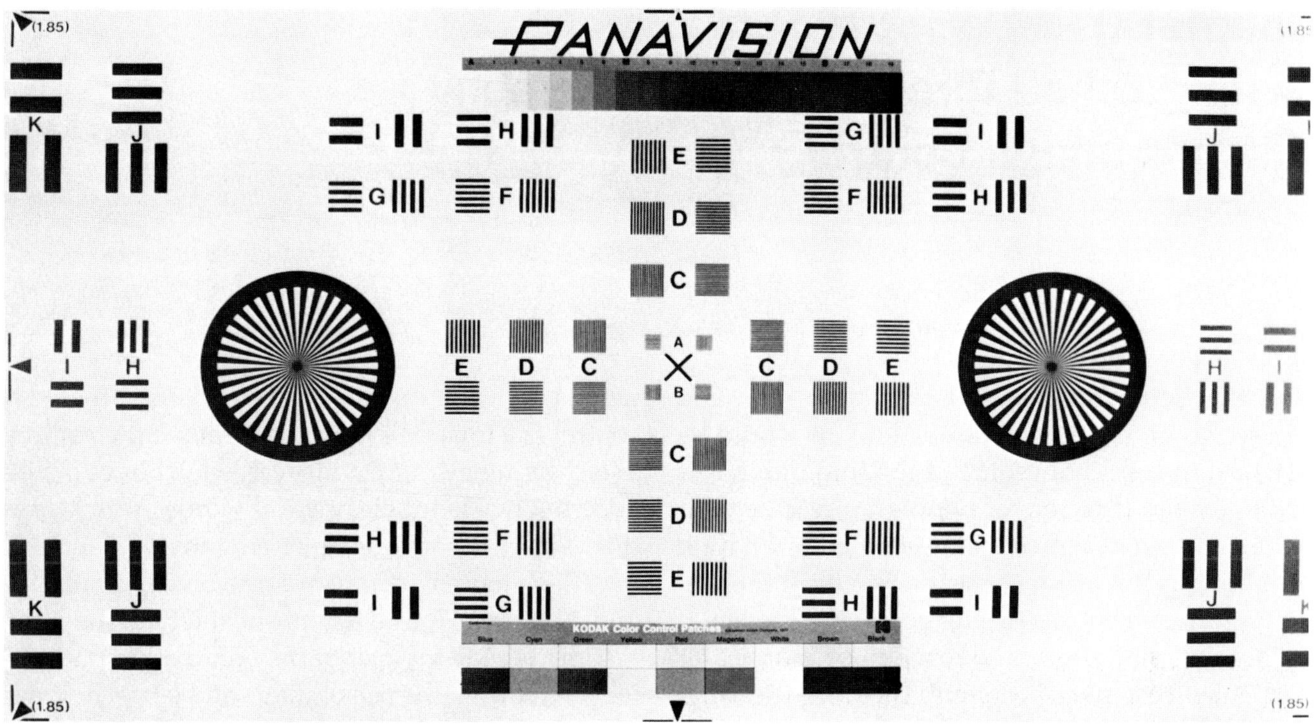

Figure 14 Lens test chart. The sharpness of a lens is evaluated by the number of lines per millimeter that the lens can faithfully reproduce, or resolve. The test chart contains a variety of equally spaced black parallel lines distributed over the film area. If a black line and the adjacent white space are sharply defined and measure 0.02mm, the resolving power of that lens is 50 lines per millimeter. The relative sharpness can be affected by the ability of the film stock to reproduce extremely fine detail; a higher EI rating usually means increased grain size, which in turn adversely affects the resolving power of that emulsion. The use of T-grains by film manufacturers has greatly reduced this problem.

son. As with lab tests, enough footage should be shot for each lens so that there's enough time to study each test on the screen. Forty-five feet of 35mm film will give you 30 seconds of screening time, which should be sufficient. (See Figure 14.)

Many directors of photography of theatrical features vary the exposure ratings for different sequences in the film. The density of the negative affects the prints and varies the look of the scenes. This relates back to Chapter 3 in which exposure was discussed as a means of altering the normal screen image. The decision to use exposure to control the screen image is not an arbitrary one. Directors of photography must have a blueprint of the overall visual impact they want the screen image to make on an audience before starting principal photography. This includes all the various controls discussed thus far.

Once you acquire a deeper understanding of exposure by working with the test material you've shot, you will see that you do have almost complete control of the film image. It's a great feeling of accomplishment when you view the completed film and see the look that you've spent many hours contemplating.

Chapter 9
Distinguished Directors of Photography Discuss Image Control

Now that you're familiar with the controls that can be used to give a film a specific look, it would be practical to learn how some of the film industry's distinguished directors of photography give their films a custom-designed look. As stated in Chapter 2, lighting is the primary means by which the cinematographer controls the image, creates the mood of the film, and affects the reaction of the theater audience. Because filters enhance the lighting, whether natural or artificial, cinematographers must first be in full command of the lighting with which they create an image before going on to augment that image.

Directors of photography use their knowledge of filters and lab procedures not only to create a story-related image but also to match scenes that have been shot on different days or in different locations. The director of photography's eye is truly the most exacting optical instrument on a motion picture set. Directors of photography must remember as closely as possible the look of one scene that will be edited next to another, which may be filmed days, weeks, or even months later. When the time comes to film the scene, the weather may be different, the schedule may not permit the film unit to wait for better conditions, the time of day may not match, and the general look of the surroundings may create a different mood. It's up to the cinematographer to make it more closely match what has already been filmed and thus to maintain the continuity of the look.

I asked several directors of photography to share some of their experiences in using camera filters and laboratory techniques. You may have seen the feature films discussed in this chapter, or you may be able to rent them on videotape so that you can see exactly what the director of photography did to control the image (within the limits of the contrast and color range of the videocassette). I have almost exclusively used their original words as written in their letters to me. As the letters came in, I realized that this chapter would become a very important one. Not only do the letters substantiate the information already given, but they offer other points of view regarding the use of filters and lab techniques. Here then are the letters from these distinguished directors of photography, listed alphabetically.

NESTOR ALMENDROS, A.S.C.

Academy Award: *Days of Heaven*
Academy Award Nominations: *Sophie's Choice*, *Blue Lagoon*, *Kramer vs. Kramer*, *Madame Rosa* (Best Foreign Film), *The Last Metro* (French César Award for Cinematography)
Films: *The Man Who Loved Women*, *My Night at Maud's*, *Places in the Heart*, *Goin' South*

I very seldom use filters, only naturally the #85 orange for outside correction of tungsten film. I don't believe in the "legitimacy" of filters. It is somehow the easiest but not the best way to work. What counts for me is what is in front of the camera, the reality of the film, not what you put on the lens. For instance, if I want fog for a particular reason in a scene, I demand fog with fog machines if nature doesn't provide. A fog filter is cheaper and quick, but does it really do the job?

Another example: If I want softness in the look of a scene, I'd rather resort to soft lighting than use a filter that cuts the "edges" of a face or a landscape.

I loathe those filters that do things that you can't reproduce by natural ways: i.e., star filters or those filters (I forget the name) that change color progressively. (Author's note: graduated color filters and color strip filters). Those stars in candle flames or lamps are sickening! Those skies that change color from top to bottom equally bother me. They fool nobody, it's obvious the camera person used a filter, audiences know.

Having said this, I will admit I have indulged sometimes in my career with filters, I'm not a saint, but I regretted it after having done it. Only for advertising and fashion do I feel more tolerant. Sorry to disappoint you.

JOHN A. ALONZO, A.S.C.

Academy Award Nomination: *Chinatown*
Films: *Cross Creek, Runaway, Norma Rae, Black Sunday, Bloody Mama, The Fortune*

Like most of our colleagues, in the beginning I used every conceivable form of filtration and lab techniques that seemed appropriate. However, I was probably following the "fads" or the "in thing" at the time. Nevertheless, the knowledge gained served an important purpose, it taught me to be more cautious and selective when altering the "look." In recent films I've gone back to the old and proven filters such as Mitchell diffusion, nets, and polarizers. The thing to consider is that the new emulsions are faster and finer grained, with greater latitude. This allows us more subtlety.

As an example, in the film *Cross Creek* we used Eastman #5293 (not now available) and exposed it at about 500 to 800 ASA using neutral densities to maintain an average T/5.6 on exteriors; our interior night scenes either studio or on location were filmed at T/3.2. We also used a medium-denier auburn color net. The print stock was Fuji because it *softened* the color contrast.

For the picture *Runaway* we used the Lightflex system together with Fuji high-speed #8514. The purpose was to be able to flash the negative as needed and on demand. We were able to introduce hues of color not found in lab control or with filtration. As you know, with the Lightflex system you expose the main subject correctly and the background slightly under, so that whatever color hue you choose won't be evident on the main subject. The use of both these tools reduced the contrast and gave the negative more sensitivity. We then printed on Eastman to *sharpen* some of the color contrast.

Filters such as the Tiffen 812 seem to dig into the shadows made by fluorescent lights. The Supafrosts, from an English company, give exterior foliage a pleasant wintery look.

Perhaps the most critical filters are the ones we must use on occasion to make a beautiful but mature leading lady more beautiful: diffusion glass and nets. That, it would seem, is always a priority. Fortunately the new film stocks and lighting materials in concert with the correct filters are there to help, but I stress testing and experimenting, always.

GERALD HIRSCHFELD, A.S.C.

Academy Award Nomination: *With These Hands* (documentary feature)
ACE Award Nomination: *The Neon Empire* (Showtime cable feature)
Films: *Young Frankenstein, Goodbye, Columbus, Fail-Safe, Two Minute Warning, To Be or Not to Be, The Car, My Favorite Year*

Fail-Safe

This film was made in 1964 and filmed on black-and-white raw stock, Eastman Super XX. The lighting was designed to create the ominous mood that permeated the entire suspenseful film. The film dealt with the accidental nuclear bombing of Moscow. The main problem arose in the large set that replicated the control room of the Omaha nerve center of the U.S. Strategic Air Force. The 18′ by 35′ screen that represented the radar images from around the world had only 4fc of incident light from the Mitchell 35mm background projector striking the latex front-projection screen. This was before Scotchlite front projection screens, before the super-high-speed films that are now available, and before we had the high-speed lenses that are now used under low light levels.

The problem was how to get a properly exposed negative under this very low light level. My solution was to overdevelop the black-and-white negative and build the gamma (contrast) of the film stock to about an 8.0 gamma. (Normal gamma for black-and-white negative stock was 6.5.) I didn't know then that what I was asking the lab to do was to "push" the negative about a stop and a half. By having the lab overdevelop the negative, I gained the film speed necessary to record the 4fc image on the screen. The extra contrast of the image, which resulted from the forced development, was used stylistically to create a graphic, almost documentary look that made the film more realistic.

Goodbye, Columbus

This was essentially a love story, and I wanted the colors to be somewhat muted and the image to have just a touch of softness. A regular fog #3 was used throughout, even on night exteriors, which gave street lamps and other lights a soft halo. Heavily diffused lighting was used to create an image softness supported by the use of the fog filter. It would be incongruous to light a scene hard and contrasty and then expect the filter to create an entirely different effect. Filters are an enhancement of the existing light situation.

This feature was filmed before there were high-speed color negative stocks, therefore I "pushed" the color negative one stop to double the ASA rating. The additional contrast caused by the forced development was offset by using the fog filter and also by using a slight increase in the exposure, about one-half stop, which not only enhanced the filter effect but also helped reduce any increase in grain caused by the forced development.

Young Frankenstein

This was another black-and-white film, and because I was satirizing the look of the original *Frankenstein*, I asked the lab to develop the negative to about a 7.2 gamma to better match the more contrasty negative used for the original. The increased development time (equivalent to forcing or pushing) gave me the desired degree of contrast and decreased the amount of light needed for the F/4.5 stop that was used to carry the depth of the large castle sets. The only filters used on this film were diffusion for the close-ups and a star filter in only one scene where the leading lady wore a spangled evening dress. I don't know if I would choose a star filter today because I think it's greatly overused.

Two-Minute Warning

The director, Larry Peerce (*Goodbye, Columbus*), asked that the film not have saturated colors, yet at the same time he didn't want as much softness as in our earlier film. Low-contrast filters were used on most of the exterior stadium scenes: everything from H&H low-con #1 to #4 and Tiffen low-con #1 to #5. The reason for this wide assortment of low-contrast filters was that very often there were three and four cameras working, and because the sun was the key light, each camera had a different degree of contrast to control because they were spread out in various parts of the stadium. By using these different low-cons, I was better able to match the image from the various camera positions and adjust the contrast from one camera setup to the next.

My Favorite Year

This was a period film, and the place was New York City. The director, Richard Benjamin, and I agreed that the feeling of 1954 would best be achieved by choosing locations that hadn't changed over the intervening years and by using clothes, hairstyles, and cars of the period. The one thing I thought wouldn't work in re-creating the period was to use the style of cinematography of the fifties. I thought it would probably look like bad photography in comparison to today's natural look.

We had to use some old footage of Peter O'Toole, our star, and we had to match some sequences to look like the old saturated color films. I decided to make the balance of the picture more contemporary or at least have a different look from the older films-within-a-film.

The scenes we shot to match into the older footage were filmed with hard light (fresnel spots) and had a more contrasty look; no filters were used. The balance of the film used soft light. The main set was the NBC studio in Radio City, New York, where soft light more closely matched the fluorescent look of the studio corridors and offices of that period. To further emphasize the difference in looks we used low-contrast filters and some double fog filters, depending on the contrast of the set colors and props. Combining these techniques gave us two very distinct looks—1954 and about 20 years earlier.

To Be or Not to Be

This too was a period film (1937), and the place was Warsaw. The first thing I learned was that some World War II black-and-white footage would be intercut with the footage that we were to shoot. At the preproduction meetings, Mel Brooks, the producer, wasn't sure where the transitions from color to black and white and back to color would be made. My suggestion was to film all the footage in color and make a black-and-white dupe negative; then, the transitions could be made anywhere, and the decision of when to make them could be deferred until post production. An additional advantage to this plan was that the ongoing action could dissolve from black and white to color within one scene (not as a transition between scenes) through an optical dissolve.

Another consideration was how to film the leading lady, who played a glamorous and mature actress, together with her suitor, a much younger man. It just wouldn't look right to use heavy diffusion on her close-ups only. Most of the scenes between them were played in her dressing room at the theater. I decided to give that set a look that would be different and also in keeping with the character of the actress. Because most of the film was shot without any special filters, it was a simple matter to make the dressing room have a much more romantic ambiance through softer lighting and a regular fog filter #3. These muted the colors somewhat and cut the sharpness very slightly. On the close-ups, depending on size, a Mitchell C or D diffuser was used on the actress along with the regular fog and sometimes a low-con for additional contrast control. For purposes of compatibility in intercutting, the young man was also slightly diffused with a Mitchell A or B plus the regular fog #3. In this way there was never an extremely sharp close-up immediately before or after one that was more diffused.

In their two-shots, which were a great challenge, I used a split diffusion filter made to order by Pancro-Mirror, which was a B-to-clear. The B side was for her, and the clear side for him. I added a Mitchell overall A diffuser, plus a H&H regular fog #3. The result was that the woman had A and B diffusion, the man only A diffusion, and they both had the fog #3.

Another control that was used to good advantage was the Lightflex system. One of the main sets was a beautiful old European-style theater with gold-leaf private boxes, a full balcony, and gilded corridors. After lighting normally, minus fill light in the background, I added the Lightflex with its candlelight filter pack, which enhanced the richness and filled in the shadow areas with a warm amber glow. Although it was a simple matter to dial in the desired degree of warm fill light for the long shots, it was another matter to judge the closer shots. There I had to match by eye and memory, dialing in more or less fill light until the shot looked just right.

The Car

I mention this film specifically for the 20-minute climactic sequence, which could be accomplished only through complex uses of film stocks, filters, and lab techniques. It was a challenge to create the desired effect. The sequence starts with the first light of dawn and progresses in time to the moment when the first streak of sunlight appears over the horizon. The sequence, which represented about 20% of the entire motion picture, took place outdoors in the rugged desert and mountains of Utah. The schedule did not permit shooting only dawn-for-dawn or dusk-for-dawn; we were required to shoot all day over a period of about two weeks. To match these very exacting scenes, different filters were used, different emulsions were used, different exposures were used, and different lab techniques were used. To learn how this was accomplished, see Table 4 in Chapter 10. *The Car* dramatically demonstrates the control that can be mastered with the application of many of the techniques described in this book.

The Neon Empire

During the preparation for this three-hour feature for Showtime, I asked the director, "What's your feeling about the look of this film?" The reply was, "Since it mostly takes place in the desert, I want it to feel and look hot!" The ball was now in my court; what was I going to do? After giving much thought to the director's request, I planned some filter tests. This was a period film covering 1934 to 1950, and I knew I didn't want the film to have the clichéd sepia-toned period look.

The locations of the film were early Las Vegas, New York City, Havana, and Los Angeles. The results of the filter tests led me to the following plan for the "hot desert" look the director wanted. For the exteriors of Las Vegas we shot on Eastman #5247 with a #85B correction filter, a coral #3, and a regular fog #3. The additional C-3 gave the film a very warm tone, and the regular fog #3 added some color muting and a very slight softening effect; together they created the desired look. For the interiors of the same location a coral #2 and a regular fog #2 replaced the C-3 and the fog #3, providing less of a "hot" feeling. Raw stock used for the interiors was Eastman #5296; the high-speed color negative gave me the ability to shoot at about F/4 to F/4.5 without requiring too much light. This enabled me to switch from primes to a zoom lens without concern for the reduced lens stop of the zoom lens.

The Los Angeles and Havana locations were filmed with a coral #2 and regular fog #2, which did not provide as hot a look as the Las Vegas exteriors. The New York City sequences were filmed with a one-half double fog and no coral filter; the one-half double fog provided just a bit of color muting to carry out the period feeling. Using these filter combinations gave the different locations a variety of looks to support the story line.

Because of the special look the director wanted, he requested that I be brought from my home in Ashland, Oregon, to Los Angeles to supervise the timing of the answer prints. It was a wise decision. Even though the camera reports listed the filters we were using, the lab completely "corrected" the "hot" feeling on the first answer print, which was made before my arrival. They probably felt that the scenes looked too yellow! On location we were sent a half-inch VHS videotape so that the director could check the work. The videotape "dailies" didn't provide me with any information about my work, so I had asked the editor to send me film clips (three or four frames) of all of the printed scenes, along with their printer lights. These later saved many hours at the lab. The timer remade an answer print based on the timing lights I had, and we again had the "hot" look we wanted.

I was also able to add my input to the film-to-tape transfer. The Rank Cintel video transfer equipment provided an additional opportunity to do some fine-honing of the image to be televised. During this process, it was a very simple matter to adjust color, contrast, and brightness by simply turning a knob. This would normally have been done by making other answer prints, but because the final U.S. release was television, this was perfect. The look of the videotape was later used to make the final answer print for the European release as a feature film.

OWEN ROIZMAN, A.S.C.

Academy Award Nominations: *The French Connection, The Exorcist, Network, Tootsie*
Films: *Three Days of the Condor, The Electric Horseman, The Stepford Wives, Taps*

The French Connection

The French Connection was a very interesting challenge photographically because the nature of the material lent itself to being done very realistically, not the usual "pretty" look that we all want to show as what we can do. For me personally that really presented some interesting decisions since it was only my second feature film and the first one hadn't as yet been released. As it turned out, the first one was never released at all!

What I decided I needed to do was to make the film look a little raw, thin, dingy, all those adjectives that are the opposite of glitsy, shiny, pretty, etc. The way I chose to do that was to force-develop the negative, underexpose it, and print it back to normal. Force-developing alone degrades the film somewhat, and if it's exposed properly, then it's not too bad. Additional underexposing on top of this, however, can really degrade the negative quite a bit. I also chose to use the old Baltar and Cooke lenses instead of Panavision because they were softer and didn't have the contrast and sharpness that Panavision lenses have. Then, of course, there was the use of the hand-held camera. We shot about 70% of the film that way. I didn't use any filters other than the #85s, as they weren't really necessary considering everything else that was being done. Sometimes when the light levels got too low for the #85, I would shoot without it in order to get the extra exposure and then have the lab correct for the missing filter. The lenses were only F/2.3 and F/2.8 so I needed all the help I could get for exposure. Also, at that time we were still using Eastman #5254 film stock, which was slower but which possessed a softer quality than the films we are accustomed to today.

Tootsie

Oddly enough, although it appears that it would be a relatively simple motion picture to photograph, *Tootsie* presented some of the most unusual problems that I have ever faced on a feature film. Everything that was done revolved around making Dustin Hoffman look believable as a woman. We felt that if the movie audience could actually believe that he looked like a woman then they would accept the fact that the characters in the film might believe it as well.

What the audience doesn't see is that it took a tremendous amount of make-up to make the transformation from male to "female." With all that make-up on for so many hours each day, it was important that we kept the heat factor on the set down to an absolute minimum. In order to achieve this there were many things to consider. Naturally, the first thing we think about when talking about heat are the lights. The smaller the lights, the less heat. However, I found out through testing that very low levels of light for some reason didn't work as well on Dustin's face as higher levels. We also were working in the anamorphic format, which I feel inherently looks better when there's a little more depth of field. Therefore, in order to satisfy both of these requirements and still keep the heat factor to a minimum, I decided to use HMI lights whenever possible as well as a faster negative stock. At that time there was only Eastman #5293 (ASA 320) available, which lacked quite a bit of color saturation compared to #5247. I decided that to keep things as consistent as possible that I would use the #5293 on both the interiors and exteriors.

Now that the heat problem was solved, I had to

deal with making Dustin look good as a woman. There were several approaches to this that had to be combined for the ultimate solution. First, he needed to be lit rather flat and soft. That in itself presented another problem because as a woman he would be wearing glasses and the flat light would constantly be reflected in them. We solved the problem by having Panavision apply a special "anti-reflection" coating to the glasses. Next, through testing, we found that he needed diffusion, but we didn't want it to be obvious, so I used a medium-density net. The flat light and the diffusion unfortunately were not enough so we discovered through some experimenting that by flashing the print and using the ENR process that Technicolor had developed we were able to smooth out all the problem areas.

GIUSEPPE ROTUNNO, A.I.C., A.S.C.*

Academy Award Nomination: *All That Jazz*
Films: *Baron Munchausen, Popeye, Carnal Knowledge, Umberto D, The Organizer, The Bible, On the Beach, Le Notti Bianche, Five Days Make a Summer*

My relationship with photography all started when, as a boy going to school, my route led me past a photographer's show window. The thing that most struck me were the blow-ups of portraits and landscapes, all enclosed in limits different from those our eye normally imposes. This excited my curiosity so that I began to limit my own vision by my hands and used my imagination to blow up what I saw. As time passed these views of mine grew from static scenes (photography) to dynamic scenes (cinematography) by aid of buses, trains, or even walking.

I then got a job in a photographer's laboratory, where, to my work of developing, printing, and retouching, I added the experience of snapping photographs with a Leica, which the lab's owner, Arturo Bragaglia, let me use every weekend. Parenthetically, the Bragaglia boys—Arturo, Anton Giulio, a theater director, and Carlo Ludovico, a motion picture director, all impassioned seekers of new photographic expressions—wrote a book on Futurist photography: *Photodynamismo*, which is still seen on view in the more important Futurist exhibitions. Anyway, this experience led me to discover the importance of light in photography and in going ahead with my career. I had a chance to put it in service in my work, developing what I had been experimenting with in the Bragaglia laboratory. Man has at his disposal natural light and artificial light, by means of which, first sunlight and then electric light, photography (writing with light) is born.

Right from my debut in cinematography I mainly sought to create atmospheres suited to the film I was shooting and to get the spectator involved, putting him in the best possible position to "get into" the story in an immediate fashion, without distracting him with light effects that were ends in themselves, and taking account of the story's setting and of the era within which the film was to be interpreted, working most particularly with the kaleidoscope that light provides us and that changes at its every smallest movement.

This kaleidoscope is mainly composed of three colors called the primary colors, of back lighting, and of diffused light (fill light). According to the distance, the inclination, the density, the contrast of these colors and lightings, and whether the light is direct, indirect, or filtered, they let me exploit a part or all of the spectrum of the radiation visible, and invisible, to our eyes.

In my first important black-and-white film *Le Notti Bianche* (literally, *White Nights*) from a short story by F. Dostoyevsky, directed by Luchino Visconti in 1957, I tackled, with the director, the film's photographic style, the necessity for a visual representation of true and false, and of reality and imagination. I thought it suitable to use two different kinds of photographic film: DuPont's for the contemporary scenes, Kodak's for those in memory. All those meteorologically distinguished

*Translated from the Italian by Linguatalent, Rome, Italy.

night atmospheres related in the story (cold nights, windy nights, foggy nights, snowy, rainy…and the arrival of dawn) were underscored using light. The fog was visualized by gathering the light, which was metered by using controlled resistances, onto huge gray veils (nets) that hung from the studio beams, dropped to the ground to divide the set into various sections while leaving the actors' paths free, these having been worked out beforehand.

Thus a particular emotion-laden vision was created, effectively bringing our dialogue between reality and illusion ahead, such as the effect created by moving the lights in harmony with the oscillation of the street lights in the wind. For the rain scenes I darkened the water in the tanks so as to make the reflections of the light through the raindrops more visible. Even the sudden arrival of the snow was underscored by large beams of light so as to make the snow appear something out of a vision, the bearer of happiness, up to the gradual arrival of dawn, in which the story ends.

To shoot *Le Notti Bianche* I used very light Kodak CC 5-10-15 filters to get a good chromatic reading of some of the colors less sensitive to black-and-white films.

In 1959, in a black-and-white film directed by Stanley Kramer entitled *On the Beach*, the need arose to intervene, not just with light in the interiors and on the sunlight outdoors with the traditional filters to blanch the verdant green of the trees and the blue of the sky and the vital rays of the sun, by placing a plate of optical glass before the objective lens, which I manipulated according to the requirements of each individual setup by tapping on it with fingertips coated with drops of liquid rubber. It was thus possible for me to break up as I pleased, depending on the filtering density obtained by the fingertip treatment, the sun's rays or the artificial light, transforming them into atomic radiations. These radiations appeared ever more intensely until they had killed the survivors on the last scrap of beach saved, up until then, from a supposed atomic war. A light but continual shaking of the camera, using a double head, put our actors in a state of uncertainty, helping the filters to convey an atmosphere of insecurity and of impotence to stop the destruction put into motion by man himself.

In setting up the photographic style of the film, *The Organizer*, I kept in mind what the mind does to our more distant memories. It tends to beautify them and to surround with affection everything belonging to the past, even the saddest events. With the aid of a few traditional filters I created a tender and almost smiling atmosphere around the protagonists and the events they lived through, even during the most dramatic scenes.

Five Days Make a Summer, directed by Fred Zinneman in color in 1981, was a complex film, an intimate love story within a Scottish family. A large part of it was filmed in the high mountains, from 6,000 to 12,000 feet up. For the part shot in Switzerland above and within the glacier crevasses, I avoided filtering out the band of rays that is invisible to our eyes but that records on film (ultraviolet light), which is strong at that altitude. I thus obtained special hues that, together with the sudden change in meteorological conditions, frequent in the high mountains, gave a taste of adventure to the couple's trip. Using light and nets I tried to give the young woman's memories a gloomy premonitory tone, limiting vision by coloring parts of the frame in black.

Baron Munchausen, shot in color in 1988, and directed by Terry Gillian, tells about a braggart, a liar, a boaster, a vain hero of amusing war adventures, and a taker of purely imaginary voyages. They are told by the novelist in a satirical and allegorical fashion. By filtering my own lights and the daylight outdoors I tried to create primary color tones, simple and reminiscent of fairy-tales, saturated hues for each of the film's episodes. For the war scenes, shot inside the walls of a city under siege, I dramatized the atmosphere in such a manner as to convey a feeling of claustrophobia that lasts up to the city's liberation.

The film *The Bible*, directed by John Huston in 1963–64, was the film that perhaps demanded the most of me in my search for visual methods to make the Bible's words concrete as it tells a story that begins before the creation. We sent a second unit around the world for six months, collecting images that could represent the creation. We got magnificent results for the creation, but we didn't have the chaos preceding creation. An idea came to me, to mix up, amalgamate, make a brew, of very cloudy skies, superimpose them all, and reverse the contrast in them. With the aid of an optical specialist we projected the various items and got a visible and formless darkness, a void in movement up to the appearance of Light. After this, the creation of man.

This was done with a stop-motion shot over several days, under identical lighting conditions. The frame consisted of a shot from a very high crane which let the camera dive toward the ground just as Adam comes alive, gets up and starts off toward the sunlight. The dust that gradually takes human form with the help of wind and a great sculptor's hands, Giacomo Manzu, is replaced toward the end by the actor. During all these takes the camera speed changed with the movement of the crane from 72 frames per second to 24 fps.

For the paradise scene and the creation of Eve, I preexposed the raw stock by exposing it to about 20% of normal exposure to a panel I had colored and specially lighted before being used on the set. This prior coloring (flashing), besides diminishing the contrast and sensitizing the negative, gave a colored background tone. It also made the image about to be superimposed on it ethereal, less corporeal, giving the paradise sequences continuity even when the lighting conditions were slightly different.

This was only the beginning of *The Bible*, but it gives you a good idea of the work required with John Huston in finding different color tones for the film's several phases. Inspiration came from the works of art of many famous painters who used different color tones to distinguish the various phases of the Bible.

In 1970 I did a film with Mike Nichols entitled *Carnal Knowledge*. In this film I had the opportunity to affect a change of photographic style within a single scene. The film begins with a night shot of an old red car parked in a courtyard with an old song as the musical commentary. A shadow comes out of the dark, near the lens. The camera begins to follow its movements until, inside the place that the music is coming from, the shadow enters the light and the face of a beautiful girl (Candice Bergen) comes into view. The camera now starts to follow her through the rooms where a university student's party is being held. I had the idea of underexposing the shot up until the girl's face comes into view so as to get into the film rather as one looks at the photos in an old photographic album; as one leafs through it, the images become clearer and clearer until you get into it and live with the protagonists.

All That Jazz, directed by Bob Fosse in 1979, contains a final dance sequence with a motif, Salute to Life, running the entire duration of the number. This sequence is accompanied by two different light atmospheres prepared under the console and used alternatively. One was created with colored lights to give the impression of life, the other with white light only. By repeatedly juxtaposing these effects an impression is given of black and white, of blood and death—a happy death though, in which we imagine we shall find what we liked in life.

VITTORIO STORARO, A.I.C., A.S.C.*

Academy Awards: *Reds, The Last Emperor, Apocalypse Now*
Films: *The Conformist, Last Tango in Paris, Tucker, One from the Heart, Ladyhawke, Peter the Great*

I often recall the emotion I first felt when lighting a stage during my early attempts at photography/cinematography. Light is energy and in this visual medium using light is an overwhelmingly sensitive experience.

I can still remember, feeling it carefully, how my whole being vibrated, molecule by molecule, in its attempt to reveal out of darkness portions of reality on a motion picture screen. It was to my eyes a reality become fantastic, this attempt to "build" from an "intuition," a way of WRITING WITH LIGHT, a way of PHOTOGRAPHING a story, a feeling, and emotion.

* Translated from the Italian by Linguatalent, Rome, Italy.

I remember vividly, intensely, how at the end of much work, of so much excitement, of so much concentration of energies within the visual space of a frame, I paused to contemplate the work done, surprising myself in thinking of how impossible the cinema is, how impossible it is to show a number of spectators that same emotion, that identically the same visual emotion that I was experiencing at that moment.

I paused thus to reflect on all the necessary, required, at times random transformations that that kind of ILLUMINATION would undergo during the whole cinematic process: various kinds of filters, lenses, negatives, developing, prints, projections, etc. That is to say, a part of the luministic-chromatic tones would be changed, however much against my will, tones that I then considered to be an integral part of a SINGLE ORIGINAL that I should never be able to put before the eyes of a spectator just as it had been originally conceived.

I firmly believed, then, that my personal, intimate expression found its completion with the light being put on the stage, that the word photography (writing with light) completed me during this first phase. Immature as my personality was at that time, I believed that the various processes that that chosen LIGHT RAY went through before reaching the screen were OBSTACLES set down in its path, to its freely being revealed.

In that innocent time of my life, I often dreamed of one day being able to directly put my personal luministic expression before the eyes of the spectators, without any artifice that might change the IDEALIZATION, the REALIZATION, the DISCLOSURE of an image.

I grasped that my true expression did not halt at the first stage of ILLUMINATION, but was completed during the whole specific voyage that had to be made right down to a specific term, a specific goal to be reached: THE CINEMA SCREEN. It was only then that I understood that what I had heretofore thought of as obstacles to the purity of expression were in fact complements to that same expression.

The path of a particular light ray possessed possibilities of modifying its natural route, possibilities made available to be used in sundry combinations, to better aid in forming a CINEMATIC IMAGE, the original fruit of a mind utilizing various processes so as to better and articulately express its personality.

In fact, a director of photography, to better tell a particular story within a particular visual expression—political-cultural-social-emotional-sentimental-etc.—uses all possible technological knowledge he has at his disposal. Those intermediate stages that the starting ray of light goes through before reaching the spectator's eye, elevates this voyage to a voyage of discovery, of culture, or expression for the viewer.

Through the choice methodology for reflecting, veiling, screening, diffracting the light of the primary source of energy, the sun;
through the choice of the kind of electric bulbs to use as sources of artificial light;
through the choice of lenses to use to gather that group of selected rays within a specific space of the frame;
through the choice of filters to suitably change, select, the wavelengths of those rays;
through the choice of possible processings of the negative emulsions;
through the choice of the kind of emulsion for that film negative that will be most suited to capturing the nascent quality of the light vibrations that have been selected up to this point;
through the choice of laboratory to see to the development, to the revelation, of this image;
through the choice of what modifications to make, if any, to their standard development process, chemical compositions, times, temperatures;
through the choice of positive film that will collect together the information gathered to this point;
through the choice of kind of balance to strike between the three lights of the primary colors (of the printers) which are to determine what we see;
through the choice of further modifications to

make the positive print, by the addition of flash technologies;

to the choice of any additional technologies to supplement the standard development of the chosen positive, such as the ENR or SCS (an addition to the already existing image of the three colors of a further image definable in black and white or super-color-saturation).

The IMAGE, the final timed print is ultimately completed by means of a type of lamp house, a type of projection lens, a type of screen, so as to be able to be presented as the summation of an ORIGINAL VISION.

A style, a language, thus ends its complex and afflicted journey, after having gotten through all those possibilities of modification that have made it unique in its quality of being the individual expression of each of the world's CINEMATIC PHOTOGRAPHY AUTHORS.

ALEX THOMSON, B.S.C.

Academy Award Nomination: *Excalibur*
B.S.C. Awards: *Eureka, Legend*
Emmy Award: *The Gold Bug*
Films: *The Sicilian, Year of the Dragon, Leviathan, Date with an Angel*

Compared to some of my colleagues I am quite sparing in the use of filters. I never, for instance, use a #85 daylight correction filter, at least not for the last 10 years. I don't think it's necessary and is correctable in the laboratory, and I personally like the "cool" look one gets by not using the #85. However, what I do do, is carry a set of coral filters (#1 through #8) because I find that it depends very much on what one is photographing, whether it be a dark wood or a row of white painted houses, as to whether it needs warming up or not. In England we are subject to such huge changes in light conditions, even during the course of one day, that it's necessary to pick a light Coral in front of the lens should one, say, start a sequence in bright overcast and find you're heading into dull overcast, or even black rain clouds and the ambiance is getting bluer.*

If it gets too black or too different then, of course, no filter will make any difference and one has to persuade the producer/director that it's necessary to stop shooting, or shoot something else that doesn't have to match anything. One doesn't always succeed of course! I remember a time on *Excalibur* when we were filming the jousting scene between Lancelot and Gawain for the honor of Guinevere when the light level got so low (we were filming with Eastman #5247—the high-speed stock had not been introduced in Europe at that time) that I was obliged to go up to John Boorman, who was both producer and director, and tell him that in my opinion we should stop shooting since my meter now read F/1.4 and that we had no depth of field with which to photograph a fight. He said, "Well, if you've got F/1.4, shoot it at that," so we shot it. On the next setup I had to tell him the meter said, "Go home," but he insisted on shooting it at F/1.4 and force-developing, which we did. I hated the results but he said afterwards that he thought the grain, lack of depth, and halation on the armour looked good, but then he had to say that because we never reshot it and it's in the picture.

It was on *Excalibur* that I first used a black dot texture screen and got to like it, and I have used them quite extensively since then. They're a curious filter; they're two pieces of glass between which has been sandwiched a sort of neutral density gelatin in which is suspended a lot of black dots which look as if they have been flicked on in an arbitrary fashion from a brush. Obviously they have flicked less dots onto the #1 filter than they have on the #5.

The reason I like them is because they are a very gentle, unobtrusive diffusion. I find I can use a #1 on a very wide long shot without any apparent softening of the image, a #5 on an extreme close-up of the leading lady, and yet make the two shots look as if they belong to the same family. There is nothing worse, in my opinion, than doing a picture

**Author's note*: As the day becomes more overcast and thus a bluer light condition prevails, a heavier coral filter will help balance the scenes.

clean, that is to say no diffusion, and then cutting to a diffused close-up to try to make an older person look better. I also find the filter doesn't halate or "blow" in the highlights as does a fog filter or a supafrost; it also doesn't "double image" like a gauze; if you hold a candle in front of it you won't get two flames like you do with a net.*

Having said all that, the black dot diffusion has a light loss of one stop, which is not all bad if you're shooting in bright sunlight or snow, but of course the converse applies if you're using them for the whole picture and you move into the studio for interiors; then you've got to find enough light for the extra stop, but in these days of almost grain-free fast emulsions, this isn't generally a problem. But, if you're going to use them, be careful; it is possible to photograph the dots! On an 18mm lens at F/5.6 or smaller stop they will show; on a 50mm at any stop you will never see them.

I used the black dot texture screen filters on *The Sicilian*, which we shot in Sicily in bright sunshine most of the time and always used neutral density filters in conjunction with the black dots so that we never shot any exteriors beyond F/6.3. The reason I did this was so that (a) I didn't photograph the "dots" and (b) so that the sky retained a natural color. I have an abhorrence of exteriors which are shot at F/16 in which the sky is so blue it looks like a traveling matte backing. I do this anyway whether I'm using diffusion or not; it also gets into the shadows more if you use a wider aperture.

To get back to the beginning, I was talking about coral filters. In my opinion, it's good that you don't just slap on a #85 because that's what the book says you have to do, but that you assess what's in front of you and decide from there. If you are shooting a dawn sequence, maybe a seascape, you have the choice of whether it's going to be a cold, blue windswept dawn or a warm, romantic affair. But, apart from the artistic evaluation, also assess the amount of blue involved, because if the reading on the blue record is too high, the print will always have a magenta cast, and that looks ugly in my opinion. I had a situation on *The Sicilian*, at the end of the picture, where the central character is waiting for a phone call at the end of a jetty with a dawn sky and the sea behind, a phone booth lit by neon, a harbor wall light which was tungsten, and a black car. I used a #3 coral and just shot it; I think it's one of the best shots I've ever done, pure Edward Hopper.

I also use neutral density filters a lot; mainly because I like to keep things back lit as much as possible so that the sky always tends to be hotter than one would like. I seem to go for soft edge neutral density grads in this case; it's tricky of course if you're going to pan and track, because somewhere along the line you're going to see the filter, but it's amazing what you can get away with using a soft edge. The only trouble is one has to push the filter down into the picture so much before the actual dark part of the filter starts to mean anything; however, I still seem to favor them rather than the hard edge NDs.

I also carry a set of coral graduated filters mostly for evening/sunset effects, although I have used them where one is looking directly out of a window into an HMI, or Brute, or whatever you're using; it takes away the ultra blue one gets in that situation.

I did a picture called *Date with an Angel* and used a white gauze on the back of the lens and a low-contrast filter on the front. This was mainly because the script called for the angel, who had descended to Earth and was unable to return, to "glow from within." The effect of the filter and gauze combination was to give her a fairly strong halation. She wore a blonde wig, had white wings, and a good strong backlight was used, which seemed to produce the effect the director was looking for. I did suggest she swallow a five-kilowatt lamp and I'd put her on a dimmer, but that idea didn't get accepted.

**Author's note*: This double image has two possible causes. When using one glass filter, some of the candlelight is reflected from the front surface of the lens (if the angle of reflection is right) to the back of the filter and then through the lens again. Panavision now makes a tilting filter holder, designed by William Fraker, A.S.C., that eliminates the problem. The double image might also occur when you're using two or more glass filters and there's space between them. Tape the filters together, eliminate this space, and the double image disappears.

I was asked to do *The Year of the Dragon* with Michael Cimino as director, which pleased me very much. After consultation we decided to force-develop the whole picture one stop whether we needed it or not so that we could get some grain and grit into the look of the picture. I was using Fuji 250 ASA, and although we didn't want to go as far as M*A*S*H, we did want it to look pretty real. We went to Technicolor in New York, since that's where we were shooting, and although they hadn't handled much Fuji film at that time, they certainly entered into the spirit of the thing, and I think the idea paid off. I also allowed the old double shadow to appear to give it greater realism and a more immediate feel; that's my excuse anyway.

Earlier this year I was engaged on a fairly big budget picture called *Leviathan*, which was made in the Cinecitta Studios in Rome. The story is about a deep-sea mining operation, and the majority of the action takes place 10,000 feet below sea level. We shot entirely on the sound stages in the studio.

On this film I didn't use any filters on the camera except for a light low-contrast (#2) on the close-ups of the two ladies in the picture. My reason was because I was using a lot of smoke, not as smoke, per se, but as atmosphere, and it wasn't necessary to diffuse the image any further. I did, however, use colored filters on the lamps and on a kind of plankton suspended in the air. I also used two different types of film stock. I used Agfa #XT-320 for the majority of the picture for all the sequences which were supposed to occur underwater and in the living quarters of the miners and Eastman #5247 for the above-water sequences. I did this to give the above and below water two different looks. I was also shooting the above-water scenes in the big tank in Malta in the Mediterranean in bright sunshine so I needed a slower negative. I also used the black dot filters for reasons already explained.

I wanted to send the "above-water" sequences to America to be processed to make the difference even more marked; the bulk of the picture was processed at Technicolor, Rome. The lab did a marvelous job, and it's no wonder that Storaro always sends his negative back there no matter where he's shooting. I have always felt that there's a difference in films that have been processed in America and those that have been done in Europe. The American result is cleaner and sharper somehow. We seem to end up with a more pastel effect. Maybe it has something to do with the water; certainly the chemical mix is the same.

I've also just completed a movie called *High Spirito*, which had an amazing amount of optical effects in it. The optical special effects supervisor, Derek Meddings, insisted on using Eastman #5247 for his work since he feels he gets the best results that way. I then decided to follow suit. It's a bit of a shock to come back to a 125 ASA negative after having done a few pictures on Fuji #8514 at ASA 500. Where one expects a F/5.6 one is suddenly at F/2.8, and that's in the studio and the director wants to use the zoom lens. Anyway, we got in a few more lamps, and I think the final result is quite pleasing. I didn't use any filters at all on this picture because the optical effects required the negative be as sharp as possible. I softened the lights rather than use anything on the lens.

Legend, however, was another matter. We used diffusion as a matter of course. It's the first time I have ever, in consultation with the director, Ridley Scott, changed diffusion from shot to shot. Ridley had the theory that you treat the film one shot at a time and not as an entity. So we would look at the content of the shot in front of us and choose a filter for that particular shot. We ranged freely from gauzes to low-contrast, supafrost, fogs, and black dots, and in the end I think it looks pretty remarkable, but you've got to be careful, and you've got to have Ridley Scott to consult with.

There have been other movies where I've tried other filters—some have been successful and others have been dreadful—but in the main, I've used filters sparingly. Thank goodness there are all sorts of combinations for us to choose from, but I think if you find something that suits you, in the main stick with it. As someone once said, there are three basic rules in the use of filters in photography; the trouble is, no one knows what they are!

Thank you for the opportunity to discuss a little

of what I think and do. I look forward to seeing what my peers have to say on this subject.

ZOLI VIDOR, A.S.C.

Distinguished director of photography of commercial films, now retired.
Winner of Atlanta Film Festival Gold Award

On several jobs I was requested to get a sepia look. Remembering some of the things I did as a member of a photo club in Vienna when we used to print on normal paper and then sepia-tone, getting very pleasing results, I planned some experimenting. I did not like the various sepia filters on the market at that time. I decided to have the set as well as the wardrobe rendered as monochromatic as possible without looking absurd. I obtained a fairly large optical flat to shoot through, set it at 45° to the lens, and reflected a well-chosen sepia or brown no-seam paper from the optical glass into the lens. By varying the light intensity on the brown paper I was able to superimpose that color over the scene being filmed through the glass, very much like the Lightflex that's being used today. Of course, the Lightflex is much more sophisticated; however, the effect I created worked out fine and was very well liked.

On another job that I recall, called *Sunrise*, everything was supposed to be filmed at sunrise. We got up at 3 A.M., filmed at El Mirage in California, freezing cold at that hour. The advertising agency loved the color of that very early light and wanted to retain that look all day. That took all kinds of doing. We used gold reflectors, gelled our lights with amber, used lots of nets to reduce the harsh sunlight, polarizers, and coral filters, and now the temperature was 120°! This was very difficult work, and all I had to go on was my memory of that very early light. This was one of the hardest jobs I can remember, both technically and physically. The job turned out very well. I received the Gold Award from the Atlanta Film Festival. I still have it!

Indoors I rarely used filters; however, there was seldom a job on which low-con, fog, star, and diffusion filters were not requested. Basically, I was always into retaining a natural look. Whatever help I needed was readily available from the multitude of filters and gels on the market. I used various split diopters, nets, and stockings for diffusion of various densities and colors.

Regarding lab instructions: Besides requesting normal or pushed processing, I always asked to have my negative printed in the middle of the printer scale so that I could tell if my exposures were off or not. If there was an extended filming period or if some extraordinary effects were requested, I always asked for a day or two for testing purposes.

VILMOS ZSIGMOND, A.S.C.

Academy Award: *Close Encounters of the Third Kind*
British Academy Award: *Deer Hunter*
Academy Award Nomination: *The River*
Films: *Deliverance, McCabe and Mrs. Miller, Obsession, Heaven's Gate*

McCabe and Mrs. Miller

I wanted to create an old, faded color picture look, much like an old autochrome transparency (by Louis Lumière) looks today. We manipulated the negative in many ways to get that look:

1. Using double fog filters (mostly #1 and #2) we created an antique, hazy, atmospheric look, especially shooting on dark, rainy days.
2. Underexposing the negative we created a certain amount of fog level.
3. Flashing the negative before developing we added another layer of fog. We flashed the negative between 10% and 25%, according to subject matter and weather conditions.
4. Pushing the film in the developer (one stop) enhanced this foggy, hazy look.

Deliverance

I had a totally opposite objective on this film, directed by John Boorman. We wanted to create a contrasty, sharp, sort of black-and-white photography look. Most of the time we achieved this look by selecting the "right" locations, shooting them at the "right" time and at the "right" weather conditions, which in this case was overcast. We waited many hours for a cloud to cover the sun, to create that beautiful green-black-gray-white canvas alongside the rivers and lakes. It made the rapids look dramatically black and white, surrounded with green foliage and a white sky.

At the beginning and at the end of the movie we were dealing with such an uncontrollable colorful world, springtime in Georgia, that we had to enlist the help of the laboratory (Technicolor's old dye transfer process) to desaturate the colors for us. They added black and white into the color prints through a black-and-white dupe negative made from the original negative. That's how we achieved that beautiful desaturated look throughout the film.

Close Encounters of the Third Kind

My biggest problem was to blend the special effects sequences with the straight production footage. Our plan with Steven Spielberg and Doug Trumbull was to deteriorate slightly the production footage by pushing the negative one stop in development to gain a little more grain and contrast. We knew that our special effects sequences were going to be shot on 65mm, but we also knew that by the time the special effects wizards were finished with a sequence (they sometimes needed three to four passes through the same negative), the look of the original 65mm film would deteriorate to the look of the pushed 35mm film.

The prediction became a reality: By watching the film from beginning to the end you can't tell the difference between straight production or the special effects footage. You see the same quality all the way throughout, which was quite an achievement in those days!

The tremendous amount of light emanating from the spaceship was achieved by an overexposure of two or three stops and using double fog filters #1 or #2.

To extend the working hours for a *dusk-for-night* shooting situation, at the Devil's Tower sequence, I used a graduated ND filter most of the time by positioning the dark section over the sky portion of the landscape.

Deer Hunter

Some sequences of the movie had to blend in with historic Vietnam War documentary footage which was grainy and contrasty. The technique here, with the cooperation of the director, Michael Cimino, was to underexpose the negative two stops and overdevelop the negative three stops! The result was not as grainy and contrasty as we envisioned, so we had to make dupe after dupe in order to achieve the documentary look.

In the earlier part of the movie I used double fog filters to create the atmosphere in and out of the steel mill, also to romanticize the look of the town and the wedding. As soon as the story took us into the middle of the Vietnam War, I pulled all the filtration to give the movie a dramatic, real, documentary look.

Obsession

This movie was divided into two sections, past and present. The present being 15 years later. To make the actors look 15 years younger I used double fog filters and occasionally nets. Then I pulled the filtration for the "now" look. I also flashed the negative more for the earlier part (15%) and less (10%) for the later. Pushing the film one stop in those days was natural for me to gain more speed (ASA 200). All of my movies at that period were shot with anamorphic zoom lenses and I needed the extra stop in exposure gained by pushing and flashing. Today I don't have to use this technique to gain speed, Eastman #5295 has an ASA 400 rating.*

**Author's note:* Since this letter was written, the popular fast film is Eastman #5296, ASA 500.

Heaven's Gate

I wanted to have a built-in hazy, atmospheric look to enhance the effect of smoke and dust used all the way through the movie. We decided to flash the negative and flash the prints also. To maintain the relatively sharp image I didn't use any diffusion filters. Flashing the negative gave more shadow detail, and flashing the print gave more highlight detail; this combination of flashing gave me the ultimate image for this movie, a richness of texture.

SUMMATION

These directors of photography offer truly golden information. The techniques vary from person to person and from film to film. Their knowledge of filters and laboratory procedures helps them tremendously when deciding about each distinctive scene and the different parts of the film that, when combined, create the whole visual experience.

Because each sequence filmed has different characteristics, you need to know exactly what can be done to solve a problem or to create an image that supports the story. Because there are no set rules, you have the freedom (within the director's vision) to do whatever you feel is best for a given situation. Your choices may be innovations that will add to the history of film.

I can't emphasize too strongly the importance of reading all kinds of books about subjects that affect the screen image. Filters and laboratory techniques are only part of the story. You need to know how a photographic emulsion is made, how it is processed, how to read a sensitometric graph, and how to interpret filter transmission charts. Study the quality and read the color content of various exterior light conditions. Read all you can about lenses, what makes them different from each other, how light is refracted and forms an image, and how to use various lighting units. In short, read everything that will enrich your thinking about what to do in the many situations you'll encounter when filming a motion picture. All of the technical knowledge in the world, however, will not be sufficient without an accompanying understanding of the screenplay and character development.

Chapter 10
Additional Applications of Image Control

The extent to which directors of photography use the different techniques available to them directly influences the final image on the screen. Even if nothing else is done to the color negative, the style of lighting, lens selection, and camera position are major factors in the look of a film. Using various filters and laboratory techniques is not a requisite for every filmed screenplay. In many situations, the natural beauty of available light is perfect for the film, and anything done to change it would be a mistake. However, with some knowledge of filters, exposure effects, and laboratory techniques, cinematographers can see what's there and make it better when surpassing reality is appropriate. Perhaps the most pragmatic value derived from this technical knowledge is the ability to match scenes when working outdoors under varying weather conditions. As has been discussed, today's tight shooting schedules prohibit production delays. In fact, it was the shooting schedule that forced us to set up an image control schedule for *The Car*.

IMAGE CONTROL SCHEDULES

We needed an image control schedule for the final sequence of *The Car* because we had to shoot throughout the entire day. The story was about a driverless car controlled by a satanic power, and the final sequence symbolized the forces of good prevailing over the forces of evil by visually progressing from darkness to the inevitable arrival of light. The final 20 minutes of the film covered the time period from the very first glimmer of dawn, when everything is almost blue-black, through the brightening of full dawn, and on into warming daylight, until finally the sun rose just above the horizon. Light conditions at the location changed so rapidly that it was practically impossible to finish even a short sequence and have the light match the next sequence. Waiting for the "magic hour"—dawn or dusk—was just not in the schedule.

I analyzed how many different periods of dawn could be distinctly defined and came up with the following:

1. Darkest dawn: Almost night, just at first light, very dark, still much black, very blue in feeling, moonlight
2. Normal early dawn: Still dark, quite blue, slightly less contrast
3. Normal dawn: Brighter than before, less blue, the soft light from the sky reduces visual contrast
4. Last of the "blue" dawn: Still brighter, less blue than before, low contrast
5. Late dawn: Dawn almost over, warming of the sky light, light is brighter, blue is gone, contrast increasing
6. Just before sunrise: Much brighter, light is more contrasty, everything warmer in tone
7. Sunrise: Full daylight with normal contrast but with the very warm look of early morning sunlight

The problem was how to film during full daylight hours, over a period of about ten days, and have the entire sequence progress through the schedule of light conditions just outlined.

Test results showed that Fuji #8517 had lower contrast and more muted colors than Eastman #5247 and so could be used advantageously for the early dawn scenes. With the cooperation of Technicolor Lab, we selected a test scene and had the timer change the color balance from normal to blue in progressive steps of four points each, going from zero to plus 24 blue. We did the same with red for the warming of early sunlight. Technicolor did more to the color balance than just adding more blue or red, but for the sake of simplicity, we agreed to call it plus blue or plus red on the camera reports. In this particular case, I intentionally did not use any color balancing filters on the camera so that the negative would retain its full color spectrum and the lab could make the necessary subtle changes in color scene by scene, as needed.

Subtle changes in contrast, however, were more easily controlled at the camera and could be varied according to daylight conditions. Low-contrast filters were used, and certain sections of the sequence were flashed 20% after exposure, prior to development, for increased softening of the early dawn effect. Exposure control was also used as a means of keeping the printer lights near midscale. Exposure varied from one and a half stops under, for the very early dawn scenes, to normal exposure for sunrise. For each scene of each day, lab instructions indicated how many additional points of blue or red were desired, so not only would the dailies show us that we were on the right track, but we'd also end up having a record for the lab for the final timing.

Polarizers and graduated neutral density filters were not listed on the image control schedule. The use of these filters was strictly dependent on the camera setup in relation to the sun, the sky, and light conditions.

The image control schedule served two very important purposes. Due to the many locations involved, we did not shoot in continuity. With the director's cooperation, scenes were rearranged even more to take advantage of the sun's position. In this way, I could have a day-for-night effect working for me whenever possible. (In this situation, it was actually "day-for-early-dawn.") Because all the scenes of this very long sequence were listed on the schedule, it was easy to know exactly where we were in terms of story progression, what the desired light effect was, what filters to use, how to change the exposure, and what to instruct the lab.

The second important use of the image control schedule was for the second camera unit, which was off by itself doing pickup shots for the same sequence. By referring to the scene number on the schedule, the second unit camera operator knew exactly what to do to ensure scene matching, and the assistant knew exactly what to write to the lab on the camera reports.

The actual schedule is shown in Table 4. It took me two days to consider all the elements and to work out all the details for this sequence. It was the climax of the film, and it had to look right. When I completed this schedule, I felt confident that no matter which scenes were filmed at what time, they would all eventually fall into proper order when edited, giving a sweeping progression of time.

You can see how the sequence progressed from "darkest dawn" to "sunrise." Near the end of the sequence, starting with scene 702, the raw stock was switched to Eastman #5247 for added color saturation. For early dawn, a coral #4 filter (C-4), which does not fully correct tungsten film to daylight, was used to maintain a colder tone. Starting with scene 673, the #85 replaced the C-4 to bring the film into normal daylight balance. Using low-contrast (LC) filters and flashing the negative of the scenes that were filmed in bright sunlight helped achieve the "flat" look of dawn. The instructions to the lab are self-explanatory. Here's a breakdown of the effects used, all of which are discussed in the preceding chapters:

- selection of raw stock
- position of camera relative to the sun
- choice of exposure
- color shifting with filters
- contrast control with filters
- negative flashing
- color shifting by the lab for special effects

This practical example shows how the many techniques discussed in this book can have a coordinated and cumulative effect on the control of dramatic images.

Table 4 Image Control Schedule: Effects, Filters, and Lab Instructions for *The Car*

Scene Numbers	*Description of Location*	*Effect*	*Filters*	*Exposure Shift*	*Lab Instructions (Effect and Color Shift)*
590–591	Garage exterior	Darkest dawn	C-4	1-1/2 stops under	Dawn (dark), +24 blue
592–599	Cross country chase	Dark dawn, sky lightens	C-4, LC #1	1-1/2 stops under	Dawn (dark), +20 blue
600–610A	Warm Spring Rd.	Slightly lighter	C-4, LC #2	1-1/2 stops under	Dawn (normal), +16 blue
611–618	Mts. of Moon	Normal dawn	C-4, LC #3	1 stop under	Dawn (normal), +16 blue
619–640	Top of Mesa	Normal dawn	C-4, LC #3	1 stop under	Dawn (normal), +16 blue
641–645	Bottom Box Canyon	Darker than normal dawn	C-4, LC #3	1-1/2 stops under	Dawn (slightly darker), +20 blue
646A–E	Top Box Canyon	Normal dawn	C-4, LC #3	1 stop under	Dawn (norm), + 16 blue
647–659	Bottom Box Canyon	Darker than normal dawn	C-4, LC #3	1-1/2 stops under	Dawn (slightly darker), + 20 blue
660–661	Top Box Canyon	Normal dawn	C-4, LC #3	1 stop under	Dawn (normal), +12 blue
662–663	Bottom Box Canyon	Normal dawn	C-4, LC #3	1 stop under	Dawn (normal), +16 blue
664–667	Top Box Canyon	Lighter than normal dawn	C-4, LC #2	3/4 stop under	Dawn (light), +8 blue
668–669	Bottom Box Canyon	Lighter than normal dawn	C-4, LC #2	3/4 stop under	Dawn (light), +12 blue
670	Top Box Canyon	Lighter than normal dawn	C-4, LC #2	1/2 stop under	Dawn (light), +8 blue
671–672	Bottom Box Canyon	Lighter than normal dawn	C-4, LC #2	1/2 stop under	Dawn (light), +8 blue
673–684	Top Box Canyon	Lighter than normal dawn	#85, LC #1	1/2 stop under	Dawn (bright), none
685–704	Top Box Canyon	Just before sunrise	#85	1/2 stop under	Just before sunrise, +8 red
705–718	Top Box Canyon	Sunrise	#85	Normal	Sunrise, +14 red

- For scenes 590 through 701, use Fuji negative type #8517, rated at ASA 80 (with C-4 or #85 filter).
- Flash scenes 611 through 663 20%.
- For scenes 702 through 718, use Eastman negative type #5247, rated at ASA 80 (with #85 filter).
- Sunrise to be fine-tuned at Technicolor Lab.

LAB TECHNIQUES AND FILTERS THAT CAN SAVE THE DAY

This section contains accounts of actual production problems that were solved through the judicious use of film lab techniques or filters.

It was late in the day, we were working outdoors, and the director wanted one last shot in the lobby of an office building with a little daylight filtering in from the dusk outside. The correct exposure required two and a half stops of light more than was available. There was no lighting equipment, and even if there had been, there was no time to set it up. What could we do?

Knowing what the film lab can and cannot do allowed us to see a perfectly good print of that scene the next day at dailies. Here's what we did. We removed the #85 filter, which gave us two-thirds of a stop more exposure, and we asked the lab to force-develop the negative one stop. Now we were only a little more than one-half stop underexposed. We asked the lab to make a normal day-effect print and to color-correct for the absence of the #85. The amount of underexposure was easily corrected by adjusting the printing about five or six points lighter. By taking these steps, there was no problem, and the director got his shot.

Here's another true story. Until Eastman introduced its new emulsion #5247, the widely used color negative was Eastman #5254, which had a pleasing degree of contrast. The new emulsion could be processed in a much shorter time and in a warmer developing bath. This was a boon to the film labs, which were swamped with so much color negative that they needed round-the-clock shifts just to process the film on time. The problem with the new emulsion was that it was much more contrasty, and cinematographers didn't like it at all. Eastman eventually corrected the emulsion and all ended well.

Why mention all this? During the transition period, when both #5254 and #5247 were in use, a producer gave me two rolls of film on location with which to shoot his commercial. One roll was #5254, the other was #5247, and the producer was unaware of the contrast difference. There was no time to send for a second matching emulsion. The solution was to expose both emulsions properly, and ask the lab to flash the #5247 by 20% before development. The flashing reduced the contrast enough so that it was impossible to tell which prints were from the #5254 and which were from the #5247!

The lab can sometimes come to the rescue when the director of photography makes a stupid mistake! Arriving back at the studio after shooting exteriors for a feature film, the assistant director pointed out to the director that, because it was 15 minutes until lunch, the crew was all there, and the camera was still set up, an insert shot that we needed could easily be made using the black asphalt of the parking lot as the background. No lights were needed, so we quickly set the camera on a tripod and made the shot, but as the assistant cameraperson was breaking down the camera and removing the gelatin filter from the internal filter slot, I noticed that it was a #85 filter. This was the correct filter for filming in daylight, but for the prior sequence I had been using a #85N6 and my meter was still set for that filter. I had just overexposed the shot by two stops! It was a dumb mistake, and because everyone had gone to lunch, we couldn't reshoot the short insert. What could we do? I made a note—a well-underlined note—on the camera report, and I called the lab and asked them to print that particular scene 16 points darker than normal. (Remember, there are eight printer lights, or points, per one stop of exposure, and we were two stops overexposed.) I told the director that there was a possibility that we'd have to redo that insert, but at dailies the next day, he asked what was wrong with the shot. It looked fine! The lab came to the rescue!

Filters, too, can play an important part in getting a shot in the can when time is a prime consideration. In the following case, if the crew didn't stop for dinner, they'd go into a penalty period, which meant extra pay for not stopping for a meal every six hours. The setting was an exterior basketball court. The sun had set, the court lights were on, and an HMI lamp was set to simulate the source of the court light. Okay to shoot? No, the director wanted the shot to be wide, and although the basketball court was dark and looked like night, the sky above the trees was too light to look like night. It would be expensive to wait for the sky to darken naturally. I used a blue B4-to-clear and a .60ND-to-clear filter. When these two filters were sandwiched together in the matte box and adjusted so that the blend of the graduates was set just below the tree tops, the sky was made bluer and darker. The clear sections of the filters didn't affect the basketball court, so no exposure compensation was needed. The sky was filtered and its brightness reduced by three stops, which was the combined filter factor. These filters made it possible to get the right shot without going into a meal penalty period.

Filters and lab techniques are valuable tools. Keep them constantly in mind, and call them into play when you find yourself in an unexpected situation.

The director of photography of a feature production on location very often has an array of filters and other special lens attachments, a sizable truck loaded with lighting and grip equipment, and a generator. This is generally true for location shoots when a call for any additional equipment would create untenable delays. With the right equipment, the director of photography can work efficiently and be supportive of the director when spontaneous situations arise. A change caused by an actor's improvisation can be as challenging as a change in the weather. Experienced directors of photography can be at their creative best when asked to do the unusual when it's least expected.

You can't be faulted for not being able to make a shot that wasn't planned, but it's much better if you can say to the director, "I know what you want, and I think I can get it with the equipment on hand." You'll feel great if you can draw on all of your stored knowledge to put the best possible image on the screen.

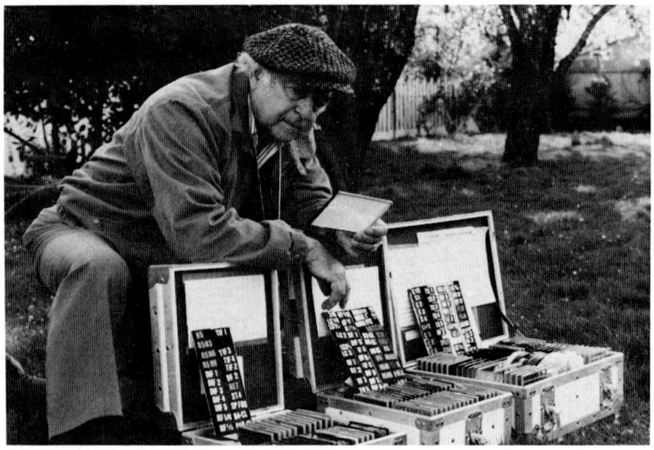

Figure 15 Jerry Hirschfeld examines a white net diffuser, one of 90 filters he carries on film assignments. Several sets of filters are duplicates, and some are triplicates for use with multiple camera set-ups. Name tabs attached to the matte box indicate the filters being used. (Photograph by Julia Tucker.)

HYPOTHETICAL PROBLEMS AND SOLUTIONS

It's good mental exercise to see how you might solve some hypothetical production problems. Someday, they could become real for you. Let's suppose you have no equipment except a camera and tripod, raw stock, and your case of filters. (See Figure 15.)

Read the following situations, figure out what you would do, and then compare your ideas to the suggested solutions. These solutions are not intended for creating special looks, just for getting the best possible screen image under what might be considered adverse conditions. At one time or another, I've faced very similar situations.

Problem 1

It's a day with blue sky and white clouds. In the scene, a car approaches and as it passes by, the camera pans with it and then holds steady as it continues driving away. The car slows as it passes the camera to show a medium close-up of the driver, who must be recognizable. All of the car windows are closed.

At the beginning and the end of the scene, the sky area is too bright compared to the lower half of the frame, creating too much contrast. It's difficult to see the driver through the windshield or side window due to cloud and sky reflections. When the car is closest to the camera, the exposure must be at least one stop more in order to see the driver, who is sitting as close as possible to the window. What would you do when the director says to shoot? (See Figure 16.)

Solution

To help take care of the brightness of the sky, you might try a graduated neutral density filter; however, this would create a problem when the car is close to the camera and the edge of the graduated filter goes through the driver's face. A better way to reduce the sky brightness is to use a one-stop attenuator, which shows no line of demarcation, reduces the exposure at the top of the frame, increases the exposure at the bottom of the frame, and is not discernible on the driver's face.

The use of a polarizer will help the problem of window reflection. Set its rotation to give you just the right amount of clarity to see the driver's face when the car passes closest to the camera, perhaps without completely losing the cloud reflection.

There are two ways to handle the exposure difference between the beginning and end of the scene and the middle close-up shot. If you have a camera with a variable shutter and a camera operator, you can change the shutter during the shot. By opening the lens one stop and closing the shutter 90° for the start of the scene, you have the same basic exposure. As the car gets closer, slowly open the shutter fully, then close it down again as the car drives away. With this method, the close-up of the car gets a full

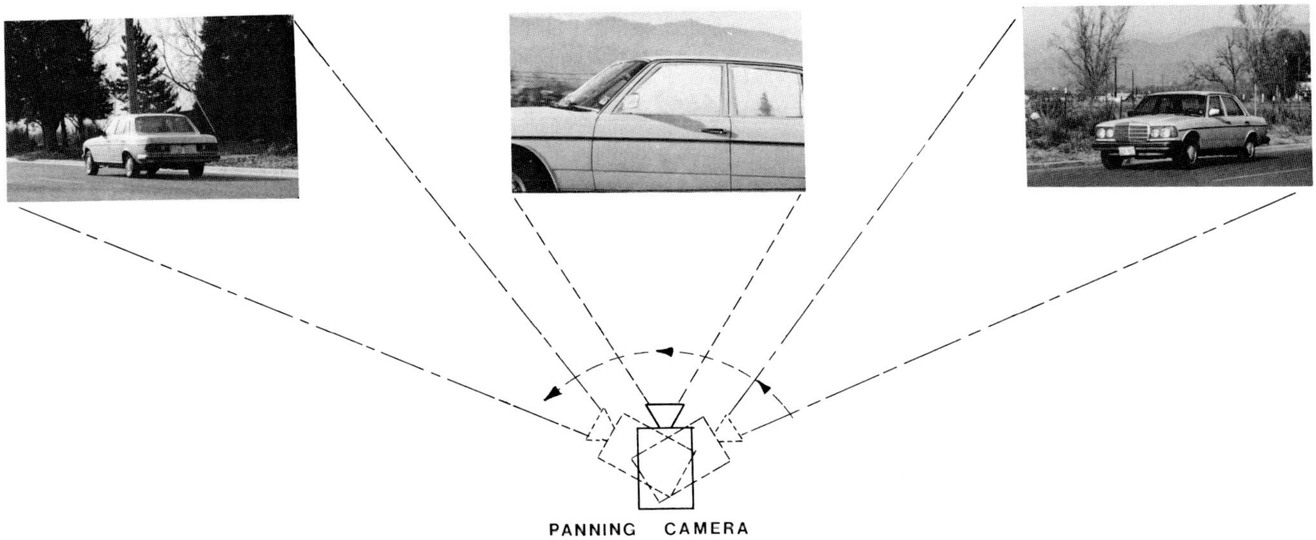

Figure 16 Hypothetical situation 1.

stop more exposure than the beginning and end of the scene. If you don't have a camera operator but do have an assistant cameraperson, he or she can slowly open and close the lens one stop to help control the exposure while also focusing the lens. (Changing exposure with the shutter is much less critical than using the lens diaphragm.)

Obviously, you must calculate your exposure compensations carefully in relation to the filters used and shutter or lens changes. Suppose you use a one-stop attenuator, a polarizer of two stops, and a shutter change of 90°: that's a four-stop change from an unfiltered scene. If the light level at your location isn't enough for a film stock with an ASA of 125, then it might be prudent to change to a high-speed raw stock, such as Eastman #5296, with an exposure index of ASA 500. Also, remember that the lab can help by force developing. (Be sure to reload the camera if this is the only pushed scene on the roll.) To control the contrast, you might normally use a low-contrast filter, but in this case you may already have three pieces of glass in front of the lens: the NBRA, the polarizer, and probably a #85. Instead of using the low con, ask the lab to give you a helping hand by flashing the negative 15% or 20%. That procedure gives you additional image density in the darker areas, especially on the driver, and doesn't affect the sky areas. With filters, varying shutter angles, raw stock, and lab alternatives, you should have no trouble getting a normal day exterior, no matter what the prevailing light conditions.

Problem 2

It's a bright sunny day, and you're asked to make a long shot through the tinted windows of an airport terminal to the tarmac beyond, where planes are lined up. The director is not sure which way would look best: (1) letting the interior foreground be semi- or full-silhouette; (2) letting the exterior go very overexposed (blown out); or (3) emphasizing both the interior and the exterior and having both areas as properly exposed as possible. In other words, you're asked to shoot the scene three ways, so the selection of the best version can be made in editing. (See Figure 17.)

Solution

For the first shot, choose a camera position that includes the best possible scene outside the windows, and keep the camera at a level that shows waiting and walking passengers in the interior foreground as silhouettes. This can be a very effective look. One-half stop overexposure for the exterior may help the bright feeling of the exterior, and be sure to instruct the lab to keep the exterior bright.

For the second shot, choose a camera position that's a raking angle, or as far from perpendicular as practical, to the windows. This accomplishes two things. First, you'll still have a sense of outside activity, but by shooting through the tinted glass at an extreme angle, you'll increase the effective density of the glass, thereby reducing the exterior brightness substantially. Second, at this raking angle, the light coming through the windows creates a cross light on the people, which accentuates them nicely. To maximize detail on the interior, expose fully for it. The windows may "blow out," or go white, as a result, but you'll still see planes. This can be a valid effect, although obviously quite the opposite from the prior silhouette effect.

To show the exterior and the interior equally well for the third shot will be more of a problem. You can keep the same angle but underexpose the people inside one-half to one stop and flash the film stock about 20%. This will give greater interior detail in the cross light, and the flashing won't alter the exterior area of the scene because it affects only the shadow areas. There's little to be gained by us-

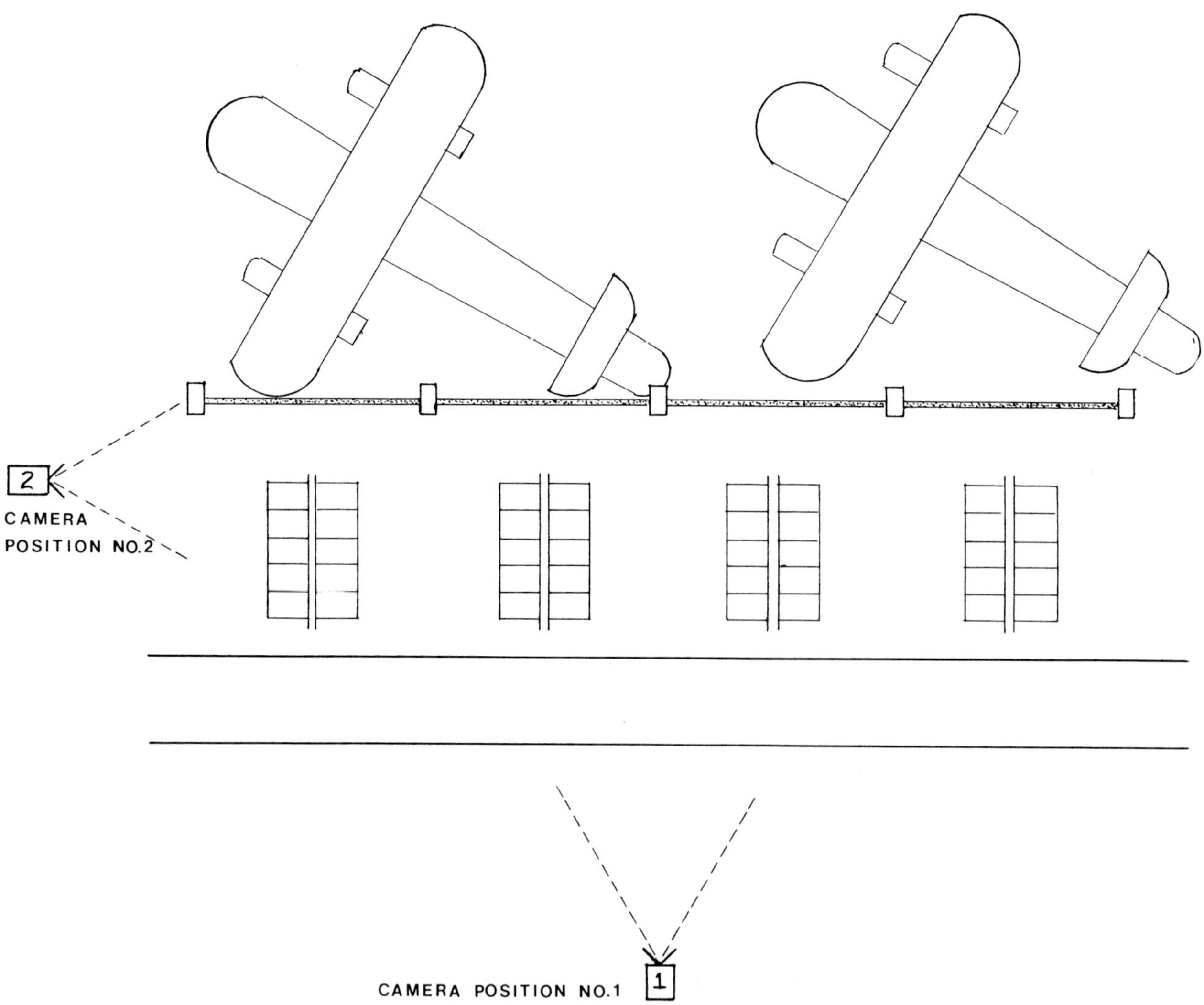

Figure 17 Hypothetical situation 2.

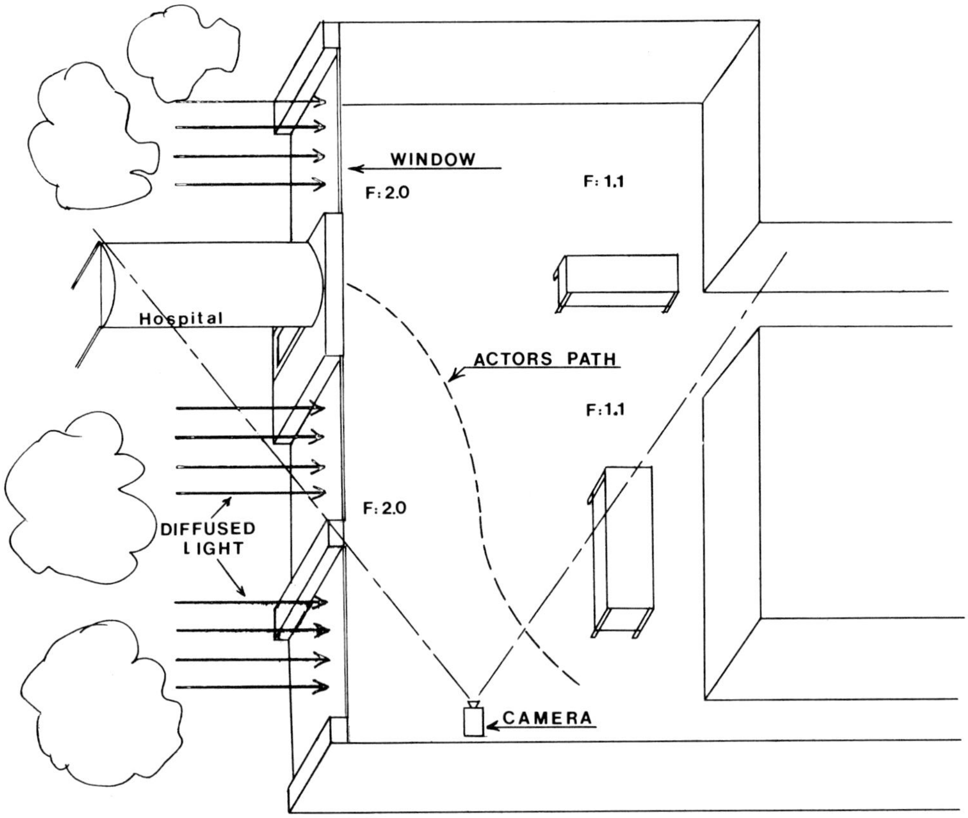

Figure 18 Hypothetical situation 3.

ing filters to control the exterior-to-interior contrast unless you're able to position a graduated ND filter just for the window area, keeping the passenger area in the clear half of the filter. A low-contrast filter would only weaken the interior subjects by reducing their overall contrast. Of course, with sufficient lighting equipment, you could balance the interior with the exterior and have the best of both worlds, but that's not what these exercises are about!

Problem 3

The scheduled shot was originally a simple one: exterior hospital, long shot, two actors walk toward the building and go inside. It was to be shot while the crew wrapped electrical and grip equipment at another location. On another day you're scheduled to shoot the continuation of the scene in the hospital lobby. Suddenly, the assistant director sees a way to combine the two shots, thereby saving a setup, so now the director wants the camera in the lobby shooting toward the doors. The actors will enter from outside and play their scene as they walk across the lobby. Because it was supposed to be a simple exterior, you have Eastman #5247 and only a zoom lens with a speed of F/2.8. The sky is overcast, and with tinted windows the effective exposure is F/2 near the windows and F/1.1 in the rear of the lobby. It's too late and too far to go for faster film stock and a faster lens. What do you do? (See Figure 18.)

Solution

The success of this shot requires the director's cooperation, and because he's the one who made the decision to move from exterior to interior, he will probably be happy to do anything he can to make your life easier! You have two requests. First, you want to have the camera at a raking angle to the wall of windows to minimize the exposure extremes between exterior and interior. Second, you want to have the background actors in specific areas: doctors and nurses dressed in white away from the windows, and others in darker wardrobe near the brighter window area. That accomplished, you can now analyze the situation and make your decisions. A basic exposure of F/1.4 (halfway between F/2 and F/1.1) would normally be correct, but the zoom lens is wide open at F/2.8, so you'd be two stops underexposed. It would be simple just to push two stops in the lab, but that would create a mismatch in quality to the rest of the film. However, a one stop push would be acceptable so you reload the camera and write instructions to the lab on the camera report. That takes care of one stop. (Camera crews always carry an extra film magazine for any contingency.)

Because you're exposing in available daylight with perhaps a little fluorescent cool white added, you'd normally use a #85 filter for color balancing. Removing the #85 filter gives you another two-thirds stop of film speed, and the last one third is easily gained by using lower printer lights. You instruct the lab; "Color correct—no #85 used—print for best possible interior." The lab will automatically adjust the printer lights in the course of making the best possible print.

Figuring the basic exposure at F/1.4 splits the difference between F/2 and F/1.1 and allows the people near the windows to be brighter than the people in the far corners of the lobby, which looks realistic to the eye. However, to the film, this difference will be intensified and appear as excessive contrast. A low-contrast filter can help solve this situation. Through the process of elimination, by eye, you might choose an H&H low-con #3, which is strong enough to help control this extreme contrast. The low-cons lend the illusion of increased exposure by dispersing light into shadow areas, so the filter will help balance the left and right sides of the frame as well as reduce the contrast on the actors' faces as they cross toward the camera into close-up and then exit the frame.

Exposure, camera position, wardrobe, filters, lab techniques, printing instructions on the camera report—it's easy to feel overwhelmed by all the necessary thought processes, but it does become easier with experience. Remember learning how to drive a manual-transmission car? At first you had to concentrate as much on shifting as on the road, but once you mastered the technique, you could just think about where you wanted to go. It's the same with film techniques.

FILTER USE FOR VIDEO RECORDING

Although enhancing the image with filters is done primarily with motion pictures made with color negative emulsions, a comparable application exists in the video medium. Assuming that you will not be using on-line production facilities, following are a few suggestions for beginning videographers. First, it's very helpful if the video camera has a manual iris control. Many of the newer models do. All have automatic irises, but when using some filters, such as low-cons or fogs, you may want to underexpose a bit to reduce the contrast without adding any milkiness to the scene. To create a stylized kind of image, such as a fantasy effect, you may want to overexpose the scene. You need a manual iris control for these kinds of exposure compensations.

Second, it's practical to use a matte box in front of the lens. Many of the camera lenses accommodate standard-size round filters, but because the lenses also have auto-focus capability, a round graduated filter attached to the lens will rotate during shooting! You may have seen the effect of a star filter attached to a lens rotating for focus. This could be avoided by switching your video camera from auto to manual focus and relying on the depth-of-field to maintain focus. If there is a focus change, the matte box separates the filter from the lens and protects them both from stray light. If you're handy with tools, you could fashion a matte box, attach it to your tripod, and use it to hold your own homemade filters. (See "Making Your Own Camera Filters" in Chapter 5.)

Third, keep in mind that the contrast range that videotape can handle is much less than that of film.

You may also find that videotape's color response is different from that of film. In view of these technical concerns, it would be wise to make tests before committing yourself to any filters. I make these same tests with my video camera before taping vacations or long trips to visit relatives.

Videographers seldom use any filters other than those built into their video cameras for color-balancing to daylight, tungsten, or fluorescent light. The reason many professionals haven't used filters more may be that they've depended on post production to create most of their effects. Camera filters can definitely be an asset to videographers, particularly when even a good post production facility cannot create what a particular filter can do. Fog filters, graduated filters, diffusers, and split dioptors are good examples.

It might be wise to explain briefly some expressions that are pertinent to video editing and the video finishing process. *Off-line editing* refers to the process in which the different scenes of the videotape are edited. This is similar to film editing except that instead of cutting and splicing scenes together, the video editor creates a log of the SMPTE time code imprinted on each frame of the video image. The SMPTE time code is imprinted on one of the video tracks and displayed on a monitor as hours, minutes, seconds, and frames. For example, 01:03:22:08 means one hour, three minutes, 22 seconds, 8 frames. The videotape is never cut; an edit decision list tells the editor exactly where one scene will end or where an effect will be placed. The time code is entered into a computer, and the electronic equipment almost instantaneously locates the next scene to be transferred to the edited tape. In this way, the various scenes are compiled in the correct order and length on an *edit master tape*. Effects may be added at this time, such as titles, dissolves, fades, sound effects, dialogue, and music, but more often, these are done during the final *on-line editing* process.

The next step to complete the editing process is going "on line." The edit master tape is transferred to the final master tape. The original material can be half-inch, three-quarter-inch, one-inch, or even the new Hi-8mm tape. Made on professional equipment, these high-quality tapes are transferred via electronic processors to the *master tape*. During the on-line process, an amazing amount of post-production work can be done. Titles may be inserted, as well as fades, dissolves, wipes, optical zooms, and superimpositions of foreground subjects on different backgrounds using the *minus-blue*, or *chroma-key*, process. In fact, almost any effect that can be imagined can be accomplished. While making the video master, an infinite number of color corrections can also be made. The image can be lightened or darkened and the contrast increased or decreased. All these possibilities may seem to obviate the need for color filters and contrast-control filters at the time of original recording. However, not all video recording is destined for a post-production facility. Many in-house production companies that make training or informational films for their own businesses never use the expensive post-production facilities; instead, they do their own simpler finishing techniques.

With burgeoning sales of video cameras and camcorders, more in-house and home videos will probably be made than were ever made on 8mm or 16mm film. Much of the information in this book about filters can be applied to this video market. This offers professionals, amateurs, and film students a new opportunity to experiment with improving their video images and effects. These videographers are not likely to visit an on-line facility where the rates for service run between $250 and $400 per hour!

The video medium has an advantage over film. The videotape can be played back immediately; you can check lighting, focus, composition, and any effect you tried to create. If you want a dramatic, deep orange sunset, you can see the results of using a #85-to-clear filter on a monitor right away. If you're presented with a very contrasty subject, you can try various low-contrast filters to see which one works best.

Very little has been written or discussed about the use of camera filters with video recording. Think of the fun you can have making up your own filter combinations and learning during the process. If your results are disappointing, just erase the tape, and try something else. Raw tape is less expensive than film, there are no laboratory processing costs, you don't need a projector and a screen, and you don't need to darken a room to see your images. Try a diffusion filter to make those close-ups look more professional.

The new video cameras and camcorders also offer professional filmmakers (directors of photography, camera assistants, and camera operators) an opportunity to experiment and increase their knowledge of filter effects. Videotape sometimes responds differently to a particular filter than film does; therefore, if you're a film person going into the video field, it's important to make your own visual tests before committing yourself to a particular filter; at least check it on a monitor while shooting.

Chapter 11
Conclusion

Summing up the experience of about 45 years brings the same kind of pleasure as seeing a completed film on a large screen. Just as a film begins with only one image, this book began with only one idea: to pass along information to the next generation of filmmakers. The negative frame is the size of a postage stamp, yet in the confines of a darkened theater, the enlarged screen image can transport an audience into worlds of fact, fiction, or fantasy. Technical knowledge gives you the power to help shape that journey.

Today's audiences are quite sophisticated about films. They can detect rear projection scenes, and they've come to expect believable location images, made possible by today's equipment and cinematic techniques. For these cogent reasons, it is crucial to stay abreast of new developments in the industry, and it is astonishing how rapidly the technology changes.

Many of the present advances in the film industry evolved because of the television industry. Film manufacturers, in an effort to create a product that television could not rival, developed high-speed color negative emulsions. With super-fast lenses, this new film enables cinematographers to shoot high-quality scenes in very low light levels—something that still is beyond the capability of professional video systems. Despite the differences between tonal ranges (140:1 for film, 40:1 for tape) there is a beautiful blend of film and video in many current productions. Science fiction films, for example, incorporate many electronically generated images, which have virtually replaced an entire repertoire of film opticals. In television commercials, as well, video graphics produce complex optical effects with ease and economy. The marriage of film and video will undoubtedly endure and produce numerous offspring.

The process of discovery has always propelled progress. Descriptions of what a filter can do are readily available from manufacturers, for example, but your own tests, your own trials and errors, can add to the general body of knowledge and satisfy your particular creative needs. In the world of film, there are constant changes and artistic challenges. When you meet these changes and challenges successfully, you'll experience an exhilarating feeling of accomplishment. When your decisions are made by dramatic instinct and confirmed by technical knowledge, you know you are doing your best, and your best can always continue to grow and mature.

No matter how many films you do, beginning a new one is the most exciting time possible: reading and rereading the script, mulling things over in your mind, exchanging visions and thoughts with the director. Finally a seed of an idea develops into a style of photography that is right for this special screenplay. Once this concept is accepted, what follows is relatively easy because the concept serves as a keystone around which all other decisions are balanced. You share with the director and the designers a common destination, and once that destination is established, together you can figure out the means that will transport you there.

At this point, theory can begin to be transformed into technique: camera, raw stock, lenses, lab, lighting, electrical and grip equipment—all these elements must be decided on now. This is the time for tests. Test the lab, the camera, lenses, film, and filters. Shoot tests for make-up, hair, and wardrobe, and possibly shoot screen tests of the actors themselves. When the test results are analyzed and the director and producer concur that the look is right, you'll have the visual foundation on which to base further decisions. Day by day, thread by thread, the fabric of the film is woven.

Watching your dailies can bring special feelings of fulfillment. You are the only person who knows whether or not what you attempted was really achieved, and you are the only person who knows what you would do the next time to make your work even better. You might make a change in light balance, use a stronger filter, a softer diffuser, whatever. You're the only one who knows what subtle enhancements are needed to perfect the image: these are things you'll include on the next picture.

After principal photography has been completed, after the editing and all the post-production work have been done, you may have the opportunity to make one more contribution: the final timing of the film. Together, you and the timer can create the finishing touches. An experienced and gifted timer is helpful, but your input is essential. Your original intentions about mood and overall look must be clear if the outcome is to match your inner vision. If the film is to be transferred to videotape for telecasting, it's even more crucial that you be there to guide the process and protect the image.

At last, the great moment arrives: the first public screening of the completed film. Not until this time will you the see the culmination of all your work. It is hoped that you will be pleased. If not, you will know what to change next time. I worked for about three years as an assistant and then as a camera operator for Leo Tover, who was a widely acclaimed director of photography for more than forty years. He told me something I'll never forget. "Jerry," he said, "the day you look at your work on the screen and say to yourself, 'that's perfect' is the day you'd better quit the business—because from then on you can only go down hill!" I strongly believe what he said because I strongly believe that you can always find something to change, to improve, to do differently on your next film.

I'd like to leave you with one more thought. Try to remember that, like any other artistic medium, film has no hard-and-fast rules. Photographic books have been written about correct light ratios of highlight to shadow; technical papers talk about correct exposures for the perfect negative; meters are designed and charts drawn up to provide the exact color temperature of a light source and explain how to use filters to match the color sensitivity of an emulsion. These are only rules—rules that can be and often are broken, but, they are broken for an artistic or a creative purpose.

Some directors of photography have developed their own basic styles of cinematography, which they may alter, depending on the type of story they're putting on the screen. Some invariably underexpose the color negative and work at the bottom edge of printer lights; others do exactly the opposite, purposely overexposing and printing their negatives in the 40 to 50 printer light range. All this is done to give the image a particular look.

There are great possibilities for individual expression in the film industry, but you must know the tools of your trade, how to use them, and how to adapt them to your own style. When and if you become an innovator, it is guaranteed that others will follow you.

Film and video are artistic mediums and subject to change. Technologies change, styles change, society changes, the world changes and if you freelance, the phone rings and the course of your life may change as you take off for a distant or adventurous assignment. Film and video stand at the center of modern history, and the director of photography stands at the center of the medium. It is a wonderful, exciting, and creative field. I am grateful for what it has given me, and I'm grateful for the opportunity of passing along my gifts and experience to you. In the future, the image will, in fact, be controlled by you.

Appendix
Filter Manufacturers

The following filter manufacturers have been producing various kinds of camera filters for years. Some manufacturers, such as Harrison & Harrison and Tiffen, have catered to the professional motion picture industry. All of the earlier filters were made of glass, either dyed glass or colored gel laminated between two pieces of optical flat glass. Today, some filter manufacturers have opted to use a new type of resin or acrylic material for their filters. There are advantages and disadvantages to both types of filters. Glass filters can crack and separate, and they are heavier and may be thicker than plastic filters; however, they are very stable and durable. Resin and acrylic filters are more sensitive to abrasion and scratches, and they require much gentler handling. This problem may be offset by their cost, which is less than that of optical flat glass filters.

The following alphabetical list of filter manufacturers is as complete as possible. It includes a partial listing of their products.

Cinemills Corporation (CMC)
3500 W. Magnolia Blvd.
Burbank, CA 91505
Tel.: (818) 843-4560
1 (800) 692-6700

CMC produces camera filters for diffusion effects. Scenes with strong highlighted areas are given emphasis by spilling light into shadows or dark areas without sacrificing image definition or focus. Very subtle effects are possible from the 10 filters in the Supafrost set. Sizes for motion picture cameras: from 138mm round to 6.6" square; for video: from series 9 to 4" square.

Cokin Creative Filter System
Cromofilter, S.A.
9 Rue De L'Annonciation
75016 Paris, France
Tel.: (33.1) 46.47.46.46

The Cokin Creative Filter System has two series of sizes—one ranging from 36mm to 62mm diameter threads; the other from 48mm to 82mm filter threads—with adapters that rotate and hold square filters. The filters are acrylic and have a wide range of effects, from solid colors to graduated. Also available are diffusers, stars, vignettes, fogs, prismatics, preshaped masks, close-up lenses, and split diopters. Most dealers have a guide book with 454 examples available.

Color Magic
1277 Leona Dr.
Beverly Hills, CA 90210
Tel.: (213) 271-6105

Color Magic will custom-make filters for professionals in sizes of 4" × 4", 4" × 5.65" (Panavision), 5" × 6", and 6.6" × 6.6". They have a quality-controlled color range of all standard colors as well as neutral density and diffusion filters.The filters are made of a highly abrasion-resistant cast resin and may be ordered as solid or graduated colors with a hard or soft edge or as colored masks. They pride themselves on 24-hour delivery in the United States and 48-hour international delivery for custom made filters. A price list and order form are available.

Eastman Kodak Company
The Motion Picture and Visual
Products Dept.
1187 Ridge Rd. West
Rochester, NY 14650
Tel.: (716) 254-1300

Eastman was the first to manufacture filters for the photographic industry and for many scientific purposes, film laboratories, and industrial printing processes. Eastman makes its filter products in either glass or gelatin sheet form; the latter can be cut to fit filter holders. Although they do not have the exotic new filters used by some, Eastman's line of filters is extensive. They are the only source of many specialty filters, and they have a full line of gelatin color-correction and color-balancing filters, which other filter manufacturers do not make. Eastman's many booklets about their filters show examples and provide transmission charts and specifications. Local dealers can be found in major national and international cities.

Fries Formatt Filters
12032 Vose St.
N. Hollywood, CA 91605
Tel.: (818) 765-3600

Fries Formatt filters are made of an extremely hard resin structure with a "glass-like" hardness. They offer a wide assortment of color-correction and color-conversion filters, as well as grads, neutrals, nets, and black-and-white contrast filters. They offer round filters in all sizes and squares and rectangular filters from 2" × 2" to 12" × 12".

Great American Market
826 N. Cole Ave.
Hollywood, CA 90038
Tel.: (213) 461-0200
Fax: (213) 461-4308

This filter manufacturer makes window gels, light diffusers, and filters for lamps under the name GamColor. Their filters are deep-dyed polyester and offer great consistency from one batch to another. GamColor filters can be used to create your own experimental camera filters. Write or call for their swatch book and the name of the nearest dealer.

Harrison & Harrison (H&H)
P.O. Box 1797 (for mail)
677 N. Plano St. (for shipping)
Porterville, CA 93258
Tel.: (209) 782-0121

H&H was for a long time the main supplier of filters for the professional motion picture industry. H&H manufactures filters in all sizes: round, square, or rectangular. H&H filters are not plastic; they are made of a special gel-like material laminated between two pieces of optical glass that are bound at the edges. Their inventory includes black-and-white photography filters, neutral density filters, color-compensating filters, combo filters, diffusers, fogs, double fogs, scenic fogs, corals, day-for-night filters, light-correction discs, polarizers, neutral blended ratio attenuators, wedges, and color-blended filters. Filters can be ordered from stock or custom-made. An informative book, *The Mystery of Filters,* available from H&H, includes descriptions of the various filters and specification charts. Mr. Harrison is also available to discuss your particular needs or photographic problems.

Hoya Corp.
Uniphot Corp.
Woodside, NY 11377
Glendale, CA 91201

This manufacturer makes a large variety of filters, multi-coated to minimize light reflections on the filter surfaces. Filters are made for color and black-and-white films. There are color conversion, color balancing, polarizing, ND, center spot, split field, fog, star, diffusers, solid and half color, and others. They make a series combining rotating polarizers and color that gives unique results. All filters are round and made in a wide range of screw-in, slip-on, bayonet, or series sizes. Ideal for still cameras and camcorders.

Note: Unable to find more information about addresses or telephone numbers. Distributed through many local camera stores.

Lee Filters, Ltd.
In the U.S.:
Lee Colortran, Inc.
West Coast:
1015 Chestnut St.
Burbank, CA 91506
Tel.: (818) 843-1200
East Coast:
23 East 22nd St.
New York, NY 10010
Tel.: (212) 995-9100

Lee got started in the film industry by manufacturing large color-correcting plastic filter sheets for use on windows of location film sets and gel sheets for use on stage and film light units. Their quality control is so excellent that they now make plastic camera filters that are more resistant to dust, water, and fingerprints than gel filters; these are used with internal camera filter holders. They also manufacture resin and polyester camera filters in a wide range of colored filters, neutral filters, and color-correction filters. A sample book of colors is available.

Lightflex (division of Arriflex Corp.)
West Coast:
600 N. Victory Blvd.
Burbank, CA 91502
Tel.: (818) 841-7070
Fax: (818) 848-4028
East Coast:
500 Route 303
Blauvelt, NY 10913
Tel.: (914) 353-1400
Fax: (914) 425-1250

Although not technically a filter, the Lightflex is mentioned here because some of its functions are similar to those of filters. There's also the new Arri VariCon, a compact contrast control unit that permits in-camera contrast reduction.

Pancro-Mirror, Inc.
6413 San Fernando Rd.
Glendale, CA 91201
Tel.: (818) 247-2238

Pancro-Mirror is a unique manufacturer of special filters for the photographic industry. They do not make colored filters, only neutral density filters and diffusion filters. Their diffusion filters are very similar to the old Mitchell diffusers, and they can be made in any degree of softness. Their neutral density filters differ from dyed or colored glass NDs in that they use a partial mirror to create the neutral density; the mirror reflects light that normally might enter the lens. They market a professional lens and filter cleaner for both glass and resin filters. They will make any type and size of filter to fit your needs. Their latest filter, Cool Mist, is similar to a double fog.

Rosco
West Coast:
1135 North Highland Ave.
Hollywood, CA 90038
Tel.: (213) 462-2233 or 1 (800) ROSCO-LA
Fax: (213) 462-3338
East Coast:
36 Bush Ave.
Port Chester, NY 10573
Tel.: (914) 937-1300 or 1 (800) ROSCO-NY

Rosco doesn't make glass or plastic camera filters. They have built their reputation on the manufacture of window gels, light gels, and diffusion materials for lamps. They were the first to manufacture correction gels for fluorescent lights, both for the tubes themselves and for the tungsten lamps that are used in conjunction with them. Rosco has a large assortment of stage gels for lights in a full range of colors. These gels can be used in front of the lens to create your own camera filters. (See Chapter 5.) Send for their sample book of Cinegels.

Schneider Corp. of America
400 Crossways Park Dr.
Woodbury, NY 11797
Tel.: (516) 496-8500

Schneider produces their filters under the trade name of B+W Filters. Filters are made of optical glass, with an anti-reflective coating front and back. They make a wide assortment of correction filters for black-and-white or color films, trick and special effects, polarizers, NDs, infrared, close-up attachments, prism filters, grads, and a lightweight front matte box. B+W Filters are made in only round sizes adaptable to lens diameters from 17.5 to 122mm. Schneider offers a very comprehensive catalog with excellent examples.

Spectratek ESS (Camera Filter Division)
7837 N. San Fernando Rd.
Sun Valley, CA 91352
Tel.: (818) 768-8064

Spectratek manufactures acrylic filters with a glass coating that makes them scratch-resistant. They make a variety of star, diffraction, and diffusion filters in all sizes, from the 52mm screw-in type to the newer 6.6" square.

Tiffen Manufacturing Corp.
90 Oser Ave.
Hauppauge, NY 11788
Tel.: (516) 273-2500 or
1 (800) 645-2522

Tiffen has recently made great strides in producing a line of filters for both still and motion pictures and for amateurs as well as professionals. Their inventory includes a wide range of filters for color and black-and-white photography. Tiffen makes a complete line of color-correction filters, color-compensating filters, neutral densities, varicolored, multi-image and vari-image lenses, star effects, color grads, fogs, low contrast, fluorescent light, vari-burst, split-field lenses, and center-spot filters. All of these are available in almost any size required. Tiffen's Photar filters are laminated filters in which the color, or other material, is an integral part of the bonding material; they are said to be equal in environmental tests to solid glass. The Tiffen net diffuser is unique in that it has actual netting material bonded between glass. Some of the net material is colored; this adds a slight color tone to the image. Tiffen has catalogs available that describe their filters and contain spectral transmission charts and specifications. They also offer the Special Effects Viewing Filter Kit, which contains 50 sample filters about 1" in diameter and a viewing filter holder so you can view by eye the effect that the filter will create.

Wilson Film Services

U.S. representative:
Levy & Associates
17 Tarleton Lane
Northport, NY 11768-2532
Tel.: (516) 269-2870
Fax: (516) 269-2871

The Supafrost diffuser has become quite popular with many directors of photography. It's an acrylic filter available in 10 densities and many sizes. Supafrost diffusers are also made with color tints: black, sepia, and blue. Supafog and Supastar, for fog and star effects, are also available.

Glossary

Academy aperture Standard size of the camera opening through which the negative is exposed in a 35mm camera. Named by the Academy of Arts and Sciences after the advent of sound on film. Prior to sound, the same opening in the camera was called *full aperture*. The Academy aperture is smaller because some of the area next to the perforations was delegated to the sound track.

answer print The first film print made from the cut negative. This print is used by the timer and the director of photography to correct color balance and print density. Several answer prints may be made until final approval is obtained regarding the look of the film. The timings of this final answer print are then used to make an interpositive and ultimately an internegative, or dupe negative. If the film was made for television transmission, the final answer print is used to make a low-contrast print for film-to-tape transfer.

arc light A source of high-intensity light equal in color balance to daylight. The light is created by a flame jumping the gap between two carbon rods connected to a direct current supply of 150 to 300 amperes. The arc light is used outdoors for location filming; however, it is rapidly being replaced by the more compact, less power-consuming HMI lamp. (See *HMI light*.)

ASA The abbreviation for the American Standards Association, commonly used as an indication of film sensitivity—for example, ASA 320—and as a reference to particular sections of the ASA standards for the film industry.

attenuator See *neutral blended ratio attenuator*.

back light The most prominent light on the subject coming from a position opposite the camera. This method is often used on exterior shots to give added dimension to the subject. The exposure is then set for the side of the subject facing the lens, or the exposure is set for the background thereby creating an effect shot (silhouette or semi-silhouette).

blue screen A filming and laboratory technique in which the subject is filmed against a special blue background using color negative film. After development, through an optical process, the foreground and background are separated to create mattes. The mattes are then used to combine the original foreground subject with any desired background without evidence of a ghost image or a fringe line around the original subject. This process is also called the *blue-matte* or *minus blue process*. In video telecasting, the process is used on news programs and called the *chroma-key process*. Some newer matte processes use a green background in place of blue.

bounce light A source of light used either as a shadowless fill light or as the main source of light. The size of the area from which light is reflected will determine the degree of softness of the shadowless light. A spot light aimed at a white wall or ceiling can fill an entire room with a soft, practically shadowless light. The light is often reflected from a 48" × 48" piece of foam core or a similar white board mounted on a century stand for a more controlled source of light. (See also *fill light*.)

century stand Also called *C-stand*. A stand made of telescoping pipes that may be raised to hold various light-controlling attachments. The top of the stand has a rotating locking device with articulated extensions to place attachments where need. (See *gobo.*)

chemical fog (1) A very slight but measurable density created on the negative during the developing process due to a reaction of the emulsion with the chemical developing bath. Chemical fog does not interfere with the quality of the negative unless it becomes excessive. (2) A laboratory technique in which the negative is immersed in a chemical foggant, which gives the negative an overall fog before normal development. This process builds density in the very thin or underexposed areas of the negative to bring them within the printing range of the print stock. The end result is similar to flashing the negative. The only lab to use this process, TVC Lab in New York, named it *Chemtone.* With the advent of the high-speed color emulsions, this process is no longer used.

color-compensating filters Filters that come in varying densities of red, green, blue, cyan, magenta, and yellow and are used to compensate for any deviation of true color reproduction. They are used primarily by still photographers to correct their color transparencies for reproduction, but cinematographers also use them to color-balance a particular lens that has a color mismatch with others being used.

color contrast The difference seen by the eye when two complementary colors are adjacent to each other, for example, a yellow or red dress against a blue wall. Color contrast may also refer to different hues of the same color. The term is sometimes used inaccurately to describe color saturation.

color-conversion filters Filters that are used to balance the type of color film to the type of light source—daylight, tungsten, or another. Conversion filters are either blue or amber and come in a number of degrees of saturation, depending on the light source to be matched. A typical filter of this type, #85 or #85B, is used to convert tungsten-balanced (3200°K) color negative to a normal daylight light source.

color correction A means of using camera filters, lab printing filters, or gel filters on lights to render a normal color image from color negative, color reversal film, or Kodachrome slides. Color correction is also the process used by film labs to change the colors of the printed scene and to match the color of adjacent edited scenes. This technique is used in the video industry for the same purpose during the on-line finishing process.

color couplers Color dyes combined with developer which, during the processing of a color negative or color print, will affect and impart color to the three individual color-sensitive emulsion layers to produce the final color negative or color print.

color interaction A phenomenon that occurs when the retina of the eye responds to adjacent colors and an effect is created that may actually alter the perception of one of the colors, even though that color remains physically unchanged. (See also *interaction of color.*)

color response The manner in which a color emulsion reacts to the different sources of light, which may vary widely over the spectral range. A color negative is said to have a normal color response when it reproduces colors accurately in the proper light source. The normal color response of a color negative may be altered through the use of colored filters or lab techniques.

color reversal internegative (CRI) A duplicating process that creates a dupe negative from the original color negative in one step. Because there is generally a slight increase in contrast of the dupe negative, the IP/IN process is preferred. (See also *IP/IN.*)

color saturation The vividness of hue. A saturated color is at its purest and deepest hue. Conversely, an unsaturated, or desaturated, color is paler. There are infinite numbers of color gradations between a saturated and an unsaturated color. Fog and low-contrast filters tend to desaturate colors.

color shift Any alteration of the normal response of a color emulsion that causes the resulting colors to be reproduced abnormally. A color shift may be controlled by a filter on the camera, by a lab technique, or by lighting conditions. Color shift due to extreme overexposure, in which only some of the colors are affected, cannot be corrected.

color temperature The rating given to a light source, designated in degrees Kelvin (°K). It is based on the temperature required to raise a black body (tungsten) to a particular color. For example, a quartz bulb is 3200°K. The color temperature of a light source can be read accurately with a color temperature meter.

color temperature meter A device that gives the Kelvin rating of the light source being measured. With this information, the cinematographer can use the proper correction filter to balance the color film to the available light source, or the lighting technician can correct the color of the light source to balance it to the film stock. Although white light is composed of red, green, and blue light, some meters read only the red/blue content of the light source. A three-color temperature meter provides the most accurate analysis. (See also *Kelvin scale*.)

composite shot The result of optically joining elements of two or more negatives in the same frame to produce a new picture. A composite shot is usually made with stationary mattes. It should not be confused with a *composite print*, on which the picture and sound track have been combined for projection, or with a *composite master*, which is the combined picture and sound track on a fine-grain positive from which the dupe negative is made. (See also *matte shot*.)

contrast ratio A measure of the range from the brightest to the darkest areas of a particular image recording medium. Color negative has a contrast ratio of about 140:1; that is, the brightest areas can be 140 times brighter than the darkest areas, and everything will be faithfully reproduced. Contrast ratio can also refer to the difference in intensity between the fill light and the key light. (See also *fill light* and *key light*.)

cool tone Also called *cold tone*. The overall effect of a color image that is more bluish than normal. This expression is often used in dialogue between the timer at the lab and the director of photography.

coral filters A series of orange-toned camera filters that range in color saturation from a very pale shade to a very dense shade of orange and are usually marked coral #1/8 to coral #14. They reduce the effective color temperature of tungsten lamps to help create the effect of candlelight or firelight, and they also decrease the color temperature of daylight. They can be used on overcast days to compensate for the bluish tinge of the overcast daylight. Corals are also used in combination with lab techniques to create a sepia-toned print because they filter out some blue and green light from the negative. A coral filter can also enhance a sunset effect. A coral #5 is equivalent to a #85 filter and can be used in its place or added to it to create the effect of a double #85 filter.

cross light The light that strikes the subject primarily from either the left or right when viewed from the camera position. This type of lighting is preferred for dramatic scenes. The subject has greater dimension than in either a back lit or front lit situation. The mood of the scene is enhanced by the degree to which the shadow side is illuminated.

dailies Also called *rushes*. The first prints from the negative that are shown to the director, the director of photography, and others to enable them to assess the previous day's work. With the advent of the VCR, dailies are often sent to distant locations to be viewed on videotape, but this is never as satisfactory as viewing the actual film footage.

day-for-night The technique used when filming an exterior nighttime scene in daylight with a variety of filters and careful positioning of the camera in relation to the sun. When properly filtered, exposed, developed, and printed, the result is a realistic night exterior scene. This technique cannot be used for scenes that include street lights and lighted storefronts.

densitometer An instrument used to measure the density of a gray scale that has been exposed and developed on a negative or print. Densitometer readings are used to evaluate the contrast of an emulsion and to check the lab's processing.

density fluctuation An aberration that is chiefly visible in medium-toned, relatively large areas of a print, such as a wall. Density fluctuation appears as a random pulse that is slightly lighter or darker than the area in which it is observable. Generally it has no regular pattern. If the density fluctuation is in the negative, there is no remedy, and the usual cause is uneven development brought about by a clogged jet port in the developing tank or the uneven addition of developer replenisher. Although this flaw is usually difficult to see in the dailies, it can show up when prints are made from a dupe negative. If the density fluctuation is not in the negative, it may be caused by a malfunctioning printer. A reprint made on a different printer usually determines if this is the case.

depth of field The distance between the nearest and the farthest objects that are in focus when the lens is focused at some point between them. Depth of field is affected by the focal length of the lens, the distance from the lens to the subject, and the lens stop used for exposure. The shorter the focal length, the farther the subject, and the smaller the lens stop, the greater the depth of field. Assistant camerapersons continually use depth of field charts to "split the focus" when subjects that must be rendered sharp are at different distances from the camera. Depth of field is sometimes mistakenly called *depth of focus*, which is an internal measurement of the distance a sensitive emulsion can be adjusted or moved to retain a sharp image.

desaturated color A color that has had its purest hues lightened by the addition of white light. Colors can be desaturated naturally by such conditions as haze, fog, or smoke, or they can be desaturated by lab techniques or fog or low-contrast filters.

desaturation technique The process used by Technicolor on such films as *Deliverance* and *Godfather II* to desaturate colors to a controllable degree. The process uses the original color negative to make a fine-grain black-and-white positive. A color interpositive is made from the original color negative. A dupe color negative is then made by printing 50% through the color interpositive and 50% through the black-and-white positive. The white light coming through the black-and-white positive mixes with the color to create a desaturated color negative. The percentages can be varied to control the degree of desaturation.

dichroic The coating on a reflective or glass surface that has the capability of selectively reflecting certain wavelengths of light while transmitting others. When used in front of a 3200°K tungsten lamp, it will transmit only the daylight (5600°K) component of the light source, converting that lamp for daylight filming.

diffraction The phenomenon that results when a point-source of light fringes into a spectrum of color at the edge of a sharp shadow. Diffraction is often confused with refraction. (See also *refraction.*)

diffused An image that has had its sharp resolution softened by an optical device in front of the lens is said to be *diffused*. Also, rays of light from the sun or a spotlight can be softened by passing through a material such as silk, spun glass, or frosted acetate that disperses the light. Sunlight is diffused by hazy, overcast conditions or by a silk or net placed overhead in a frame.

diffusion (1) A filter made of glass, plastic, netting, or other material used to soften an image rendered by a very sharp lens. It does this by refracting a portion of the light passing through the lens. Diffusion filters are generally used on close-ups because they soften imperfections and facial lines. Some diffusion filters affect only the sharpness of the image; others also reduce the contrast of the image, thus increasing the softening effect. (2) A material used in front of a light source to disperse and control the otherwise sharp light rays.

diopter A supplementary lens used to shift the normal close-focus to a much closer focus. The extent of the change depends on the power of the diopter. Diopters are often used with a zoom lens, which has a limited close-focus calibration, to allow the lens to be used at a closer distance. (See also *split diopter*.)

distortion Applies to many facets of photography where the normal image is changed so that it no longer truly represents the original. In reference to a lens, distortion is the alteration of normal perspective; to a filter, it may mean changing an otherwise normal image; to scene composition, it means using the camera at extreme angles; and in film processing, it means a print quality other than normal.

double fog filter Similar in effect to the regular fog filter, but it does not affect the resolution of the lens or create a halo around bright lights or bright areas. Double fog filters are rated according to their density and effect, that is, double fog #1/4 to double fog #5. They are often used to reduce contrast and to desaturate colors. (See also *fog filter*.)

dub (1) In sound recording, dubbing is the process of adding additional sound effects or voices to an existing sound track. (2) In the video field, a dub is a copy of the finished video product, similar to a print of a film. It can be made in half-inch, three-quarter-inch or one-inch formats.

dub master The copy, or duplicate, of the original master videotape made during the editing process of a video production. The dub master protects the original master tape and is comparable to the dupe negative of film. Any number of copies in any format can be made from the dub master. When used with digital recording, no loss of quality occurs through any of the finishing processes. (See also *master tapes*.)

dupe negative Short for *duplicate negative*, it refers to the copy of the original negative. There are three reasons for making a dupe negative: (1) to combine optical effects with the original negative; (2) to make one hundred or more prints for distribution, thereby protecting the original from any possible damage or wear in the printing machines; (3) to send the negative overseas for printing, thus saving the cost of shipping many prints.

emulsion In photography, the light-sensitive coating on cellulose acetate used to create negative and print materials. A black-and-white film has only one light-sensitive layer of emulsion. A color negative has three layers, each of which is sensitive to a primary color.

ENR process A special lab technique applied only to release prints that involves developing the print in both a color developer and a black-and-white developer. The process creates more saturated colors and exceptional blacks. Technicolor's laboratory created this process.

exposure index (EI) A classification of the light sensitivity of the negative material. It is similar to an ASA rating.

fill light The light that is generally used to partially illuminate the black shadows caused by a strong main light source, whether indoors or outside in sunlight. The most pleasing fill lights are those that do not create a shadow of their own; therefore, bounce lights or heavily diffused lamps are often favored as fill lights. The intensity of the fill light, when compared to the intensity of the key light, creates a lighting ratio that helps determine the contrast of the scene.

filter factor The number by which the exposure is multiplied to compensate for the light lost as a result of using a filter. A filter factor of 2 means that the cinematographer must increase the exposure two times, or one full stop. A factor of 4 means increasing the exposure four times, or two full stops.

flashback A cinematic technique used to illustrate a period of time before the present. Flashbacks often employ an optical device or color shift to signal the change and to separate it visually from the rest of the film. In a color film, the flashback scene may be shown in sepia or black and white or with an optical effect such as a vignette.

flashing A laboratory technique in which the color negative or positive is exposed to a very weak source of light either before or after shooting or before printing. (See Chapter 6.) Generally, flashing is used to control the scene contrast and to help see detail in the dark shadow areas.

fog filter Originally used to help create a fog effect in theatrical films, fog filters are glass filters with varying degrees of milkiness used to create the illusion of very light to very heavy fog. When used in front of a lens, the fog filter disperses a set amount of white light onto the negative and reduces both the visual contrast and the color saturation of the scene. Regular fog filters affect both shadow and highlight areas of the scene, and they create halos around very bright lights, much the same as real fog will. They also soften the resolution of the lens slightly. (See also *double fog filter*.)

Foley effects Sound effects such as creaking chairs, gun shots, applause, wind, and footsteps, created by Foley artists, who are specialists in producing realistic effects that were not originally (or improperly) recorded on the soundtrack. Other sound effects may be taken from a sound effects stock library.

forced developing Also called *pushing* or *push processing*. A laboratory procedure used when a negative is purposely underexposed by one or two stops. The negative is developed for a longer than normal time, resulting in a negative of relatively normal density. Because different emulsions react in different ways to forced development, it is essential that the cinematographer make a definitive test before using this technique.

front light The light that strikes the subject fairly head-on from the camera position. This type of lighting produces the purest colors; however, the picture can seem flat and actors may have difficulty with the light directly in their eyes, particularly in direct sunlight. Front light is commonly used on television sitcoms; it helps when using multiple camera setups.

gamma A word used to specify the degree of contrast that results from the developing process. In black-and-white photography, the cinematographer can request that the negative be processed to a certain gamma, which allows the contrast of the image to be controlled. Gamma designation is used by film labs to check the standards of their processing for both color and black-and-white film.

gobo A device used to shield the lens from a light source, or to block unwanted light from a set or an actor. Gobos are made of flat-black fabric in a variety of sizes, generally held in place by a century stand.

graduated filter A filter used for a specific purpose. One half of a graduated filter is usually a neutral density or a certain color, and the other half is generally clear. When used to control brightness, the neutral density half is used on the top for sky or on the bottom for snow or sand. Grads are often made to order in a great variety of colors for specific effects.

grain The granular texture visible in processed photographic materials. Grain is usually mentioned in relation to the negative image. A grainy negative is one in which the particles making up the image are easily visible in the projected print as "dancing" salt-and-pepper granules. In a normal-speed color negative or a high-speed color negative, grain is not a problem unless the processing is forced. New developments in the structure of the grains used to make up the emulsions (T-grains) have greatly lessened the problem of grain.

griffolyn A smooth white-surfaced material that is stretched on a frame to reflect (bounce) sunlight or tungsten light onto the subject being filmed. Its light is virtually shadowless.

high key A style of lighting where the shadows are illuminated and all areas of the film frame are brightly lighted. High key was used in many musicals and comedies of the forties and fifties. Today, many television sitcoms use high key lighting because of multiple cameras used at angles that may show the shadow side of a subject. This style of lighting accommodates the reduced contrast ratio of the video system.

HMI light The abbreviation for Halogen-Metal-Iodine light. It offers a high-intensity daylight-balanced light output. The light source is a special bulb enclosing gas, an arc discharged by alternating current, and high voltage from a ballast. HMI lamps range from 575-watt to 18,000-

watt units. On many locations they replace the bulkier, heavier arc lamps.

interaction of color An apparent change in one color when it is adjacent to or surrounded by another color. A similar reaction occurs within black-and-white tonal ranges: a subject can appear light against a dark background and dark against a light background. Interaction of color also occurs when one saturated color is backed by a complementary saturated color: the image appears to be slightly out-of-focus when in reality it's sharp. (See also *color interaction.*)

IP/IN Abbreviations for interpositive and internegative, which are film stocks of special quality used in duplication. An interpositive, a low-contrast print, is made from the original color negative. This print itself is then printed as if it were a negative, and the result is an internegative, which is also called a *dupe negative.*

Kelvin scale A system of measurement devised by Lord Kelvin of England that is currently used to identify a particular color of light. When a black body (tungsten filament) is heated to a specific temperature, it begins to glow: first dull red, then bright red, and eventually white. The standard quartz bulb used for color filming is rated 3200°K, and average daylight is rated 5600°K.

key light The main light used to establish a source of light on the subject being filmed. The sun is the key light outdoors. Indoors, the director of photography decides where to place the main light source. With an interior day scene, a window is the logical source, supplemented by lighting units that serve as the key lights.

light-balancing filters Filters of two different colors: bluish and yellowish. Blue light-balancing filters are used to keep the color film emulsion in balance when the light source becomes redder than normal, such as near the end of the day. Yellow filters are used when the light source is bluer than normal, such as on an overcast day. The density of the filter selected depends on the degree of correction needed. These filters are used mostly for 16mm reversal films and color positive slides.

Lightflex A matte box used in front of the camera lens that incorporates a specially designed, curved optical glass through which the film is shot. The Lightflex reflects the light created in a lightbox at the top of the device, which can be filtered for color or can be white light. The degree of intensity can be accurately controlled by an electronic dimmer. The device primarily effects the shadow and low-density areas, not the fully exposed areas. It is currently being replaced by the Arri VariCon device, a much more compact unit. (See also *VariCon.*)

liquid gate See *wet gate.*

look of a film The visual impact of the photographic style created by the director of photography. The look helps set the cinematic mood that supports the story. It is primarily established through the style of lighting and is further enhanced through the use of filters and lab techniques.

low-contrast filter A filter that helps do exactly what its name implies: it reduces the image contrast. Low-contrast filters affect the shadow areas, not the highlight areas. Fog filters, which also reduce contrast, affect both areas.

low key Lighting that creates a very dramatic style with heavy or black shadows. This lighting style is often used in the genre of heavy drama for mystery and horror films and in many night scenes.

master tape Usually a one-inch videotape that is the final product of all the editing, optical effects, titles, and corrections necessary for the completion of the video production. The master tape is comparable to the final edited and cut original negative of motion picture film.

matte shot Similar to a composite shot in that two or more elements are combined to create a new picture. A matte shot, however, uses a black mask to protect part of the frame while the other part is exposed. A second matte is then used to cover the exposed area while the second exposure is made. The blue-screen process is often the beginning of a matte shot in which a foreground subject is to be superimposed on a different background; this process uses the blue background to create a matte that moves with the subject and is called a *traveling matte.* A matte can either be stationary, created by a matte artist, or a traveling or moving matte created by the optical department. (See also *composite shot.*)

muted color A color with a reduced hue, or desaturated color, that has been altered by means of a camera filter or a laboratory process that adds a calculated degree of white light to all colors. If colors are muted extensively, they become pastels.

negative density The degree to which a negative, upon development, has achieved a visual record of its exposure to light. A negative of normal density, whether black and white or color, will print in the middle brightness range of the lab's printing equipment. An underexposed negative is said to have a "thin" density, and conversely, an overexposed negative has a "heavy" density.

netting A type of material that, when used in front of the lens, acts as a diffuser and sometimes imparts a color to the recorded image.

neutral blended ratio attenuator (NBRA) Commonly called an *attenuator,* it is similar to a graduated neutral density filter but has no blend line. The density change starts at one edge of the filter and gradually decreases to zero at the opposite edge.

neutral density filter A filter with a colorless gray tone that is primarily used for exposure control. A neutral density (ND) filter is graded or marked with every 0.30 of density equal to one stop. The ND filter is often manufactured in combination with other commonly used filters, for example, #85N3 and #85BN6. When the neutral density is graduated to clear, the filter is generally used to control sky brightness.

off line The process in which unedited videotape is imprinted electronically with a time code denoting hour, minute, second, and frame. The time code is used to identify a scene for editorial purposes. The off-line editor notes the exact time code at which a cut, or other effect, is to be made. (See also *on line.*)

on line The next step after completing off line. During the on-line process, the time code information helps the on-line editor determine which sections of video-tape are to be used and where. All optical effects, titles, and any necessary color and brightness adjustments are made during the on-line process, and everything is transferred to the master tape. (See also *master tape* and *off line.*)

optical flat A glass that is free of any color or optical aberrations. Normal glass has a slight green tint. Water white glass is absolutely colorless and although not optical, is fine enough for making experimental filters.

opticals Effects that are added to the original negative during post production, including dissolves, fades, wipes, blow-ups, titles and subtitles. More involved optical effects include various types of matte work, blue-screen composites, optical zooms, and transitions from color to black and white or vice versa. The resulting negative that includes the optical effects is always a duplicate negative.

polarizing filter Neutral in color, thereby not affecting the color of any subject, it has two principal functions: to reduce or eliminate undesired reflections from surfaces such as glass or water and to darken a blue sky and accentuate clouds.

post production Any work done after the completion of principal photography of the film. Post production can include second unit filming, voice dubbing, editing, optical work, titles, scoring, and any other treatment of the film before its release.

printer light number A designation of the intensity of the light emitted from the printing machine when making a print from a negative. In color printing, three printer lights are used for each scene, representing the intensity of the magenta, cyan, and yellow light. These lights are used additively to print from the color negative.

pull processing A lab technique in which the negative is developed for less than the recommended length of time. Pull processing is only done when the exposure has been intentionally increased one or more stops above normal. The results are finer grain and a reduction of contrast. Thus, pull processing may be helpful when extensive optical work is required.

push processing See *forced developing*.

raw stock Unexposed photographic film used for making negatives, prints, and opticals.

refraction The bending of a light ray as it travels from one transparent medium through another of different density. It is this phenomenon that allows a lens to work. A light beam is refracted once when it hits the curved front surface of a lens and again as it exits from the rear of the lens to the air. Lenses are constructed so that they gather light rays and refract them to form an image. Similarly, a mirage is formed when light from the sky is refracted as it strikes the extremely hot air immediately above an asphalt road or desert sand.

resolution Also known as resolving power. The ability of a lens or a film emulsion to record very fine detail. The measurement is based on photographing a series of thin parallel black lines separated by white spaces of equal width. The result of the measurement is noted as the maximum number of lines per millimeter that the lens or film can record. The resolution of a lens can be affected by anything in front of the lens that is used to modify the image, for example, a filter. The resolution of a negative can be affected by processing. For example, forced development may increase the grain so much that it affects the sharp detail. (See Figure 14.)

saturated color See *color saturation*.

sepia tone A reddish brown color often associated with an old-fashioned or period look. It can be achieved by labs with controlled color printing; however, many cinematographers prefer to put the desired effect on the original negative and to have the lab intensify that effect if needed. The use of a coral filter with some consideration for the filter factor often gives a good sepia effect. (See Color Plate 12.)

sky filter Any filter used to enhance the color of blue sky and to delineate clouds, for example, ultraviolet (UV) filters and polarizers.

SMPTE Abbreviation for the Society of Motion Picture and Television Engineers, which has set many standards for the film and video industry. The SMPTE (pronounced "Simptee") code is a time code that indicates real time: hour, minute, second, and frame number. It runs continuously throughout a 24-hour period and is used for video editing.

soft light An artificial light source that, because of its broad, white, reflective surface, creates a soft-edged, almost shadowless light. Soft light also refers to sunlight filtered through a silk-like material, which spreads the spectral source in such a way that it no longer creates a sharp shadow. (See also *bounce light*.)

split diopter Exactly what its name implies: a diopter that's been cut in half, or made originally as a semi-circle. These optical glasses are used to create sharp images when both foreground and background subjects are within the frame and the depth of field cannot carry the focus to both subjects. (See Figure 9.)

splitting focus The technique of focusing somewhere between the nearest and the farthest objects the cinematographer wants to be sharp. Splitting focus works hand in hand with depth of field and is contingent on the same principles. (See also *depth of field*.)

star filter A distortion filter that refracts a point of light into a star burst. The effect depends on the intensity of the point source of light, its contrast to the background, and the strength of the filter. Multipointed stars are created by the number, angle, and closeness of engraved lines.

strip filter A glass or plastic filter that usually has a strip of color of about one-half to one inch in the middle. The balance of the filter is clear. If the strip is blue, it can be used to increase the color of the sky near the horizon where the blue sky light usually appears much lighter. An orange strip filter can be used to enhance the horizon area in a sunset scene or to create a warm glow along the horizon in a day scene. Care should be taken that actors do not cross into the strip area.

timing One of the last procedures that a film goes through. The timer in the lab works with the director of photography to make scene-by-scene corrections for color and print density so that all the desired effects, moods of lighting, and details of visual continuity are finalized.

VariCon A device made by the Arriflex Corporation that has the same functions as the Lightflex and is similarly controlled by an electronic dimmer. The VariCon is much smaller, only about 6.5" × 2". The VariCon fits between the camera lens and the matte box; it snaps into a 5" × 6" Arri matte box but can be used with any large matte box. (See also *Lightflex.*)

videographer The person handling a video camera in any recording situation, from news coverage to major productions.

vignette The center area of a film frame, usually round or oval, with borders that are indefinite and blend softly outward to the edges of the frame. The edges are shaded in either neutral density or white. The vignette effect focuses attention on the person or scene in the center and is used often to isolate flashback scenes and sometimes to indicate a dream sequence.

warm tone The opposite of *cool tone.* The look of a scene when an extra degree of red, yellow or brown is added, either through filters or lab printing techniques. Filming outdoors during early morning or late afternoon, when sunlight is redder than normal, is another way to achieve warm tones.

wet gate Also known as *liquid gate.* A lab process in which the film is coated as it passes through the printer gate with a special liquid that reduces many abrasions and scratches. The liquid is immediately removed by an air squeegee. The wet gate is used when transferring a low-contrast print to tape to produce a cleaner, sharper image.

Index